A Manual for
Biomaterials/
Scaffold Fabrication
Technology

MANUALS IN BIOMEDICAL RESEARCH

Series Editor: Jan-Thorsten Schantz (*National University of Singapore, Singapore*)

Published

Vol. 1: A Manual for Primary Human Cell Culture
by Jan-Thorsten Schantz and Ng Kee Woei

Vol. 2: Techniques in Microscopy for Biomedical Applications
by Terje Dokland, Dietmar Werner Hutmacher, Mary Mah-Lee Ng and Jan-Thorsten Schantz

Vol. 3: A Manual for Biochemistry Protocols
by Markus R. Wenk and Aaron Zefrin Fernandis

Vol. 4: A Manual for Biomaterials/Scaffold Fabrication Technology
by Gilson Khang et al.

Forthcoming

Vol. 5: A Manual for Laboratory Animal Management in Biomedical Research
by Jonathan Ward et al.

Vol. 6: A Manual for Intellectual Property Management
by Maik Brinkmann et al.

Gilson Khang
Chonbuk National University, Korea

Moon Suk Kim
Korea Research Institute of Chemical Technology, Korea

Hai Bang Lee
Korea Research Institute of Chemical Technology, Korea

Manuals in Biomedical Research – Vol. 4

A Manual for Biomaterials/ Scaffold Fabrication Technology

World Scientific

EW JERSEY • LONDON • SINGAPORE • BEIJING • SHANGHAI • HONG KONG • TAIPEI • CHENNAI

Published by

World Scientific Publishing Co. Pte. Ltd.

5 Toh Tuck Link, Singapore 596224

USA office: 27 Warren Street, Suite 401-402, Hackensack, NJ 07601

UK office: 57 Shelton Street, Covent Garden, London WC2H 9HE

British Library Cataloguing-in-Publication Data
A catalogue record for this book is available from the British Library.

Manuals in Biomedical Research — Vol. 4
A MANUAL FOR BIOMATERIALS/SCAFFOLD FABRICATION
TECHNOLOGY

ISBN-13 978-981-270-595-2 (pbk)
ISBN-10 981-270-595-3 (pbk)

Typeset by Stallion Press
Email: enquiries@stallionpress.com

Printed in Singapore.

Preface

In the May 22, 2000 issue of the popular *Time* magazine, it was predicted that tissue engineers would have the hottest job in the new millennium, with drug designers coming in third. Indeed, the 21st century has opened a new era for the production of artificial organs by means of tissue engineering and regenerative medicine (TERM) to repair or replace damaged/diseased tissues and organs. With an increase in the average age of the population as well as in the incidence of age-related "wear-and-tear" conditions and traumatic injuries/diseases, the shortage of healthy donor organs has led to the emergence of TERM.

To reconstruct a new tissue by tissue engineering, triad components are needed: (1) cells, (2) biomaterials, and (3) bioactive molecules. Of these three components, scaffolds play a critical role in the reorganisation of neotissues and neo-organs. Scaffold matrices can be used to achieve cell delivery with high loading and efficiency to specific sites. The manufacturing methods are very important for the specific organs because the physicochemical properties of scaffold matrices — such as porosity, pore diameter, and specific area — are determined by the manufacturing methods. This book focuses on 21 different types of manufacturing protocols for tissue-engineered scaffolds that are adapted for the undergraduate and graduate student level.

We would like to especially thank Professors Jan-Thorsten Schantz and Dietmar W. Hutmacher at the Tissue Engineering Laboratory, National University of Singapore, for recommending us to edit this manual. Special thanks go to Kim-Wei Lee and Wanda Tan for their help in editing this book. We hope that this book will be very useful for students and scientists in academia and industry in the TERM field.

G. Khang, M. S. Kim & H. B. Lee
December 2006

Contents

List of Figures

List of Contributors

Eunhee Bae, PhD
REGEN Biotech, Inc.
Sungnam 462-120
Korea

Jin Woo Bae, MS
Department of Molecular Science & Technology
Ajou University
Yeongtong, Suwon 443-749
Korea

Young Ju Choi, MS
Laboratory of Biomaterials
R&D Institute, MCTT
Nowon-ku, Seoul 139-706
Korea

Mark van Dyke, PhD
Wake Forest Institute for Regenerative Medicine
Wake Forest University Health Sciences
Medical Center Boulevard, Winston-Salem
NC 27157
USA

Gilson Khang, PhD
BK-21 Polymer BIN Fusion Research Team
Chonbuk National University
Dukjin, Jeonju 561-756
Korea

Myung Seob Khil, PhD
Center for Healthcare Technology Development
Chonbuk National University
Dukjin, Jeonju 561-756
Korea

Byung-Soo Kim, PhD
Department of Bioengineering
College of Engineering
Hanyang University
Seongdong-ku, Seoul 133-791
Korea

Chun Ho Kim, PhD
Laboratory of Tissue Engineering
Korea Institute of Radiological and Medical Sciences
Nowon-ku, Seoul 139-706
Korea

Hak Yong Kim, PhD
Department of Textile Materials Engineering
Chonbuk National University
Dukjin, Jeonju 561-756
Korea

Hyun Do Kim, MS
Polymer Science and Engineering
SungKyunKwan University
Jangan-ku, Suwon 440-746
Korea

Hyun-Man Kim, PhD, DDS
Department of CMF Cell and Developmental Biology
School of Dentistry
Seoul National University
Chongro-ku, Seoul 110-460
Korea

Moon Suk Kim, PhD
Medical Science Division
Korea Research Institute of Chemical Technology
P.O. Box 107, Yuseong, Daejeon 305-600
Korea

Sang-Soo Kim, PhD
Department of Bioengineering
College of Engineering
Hanyang University
Seongdong-ku, Seoul 133-791
Korea

Soo Hyun Kim, PhD
Biomaterials Research Center
Korea Institute of Science and Technology
P.O. Box 131, Cheongryang, Seoul 130-650
Korea

Soon Hee Kim, MS
BK-21 Polymer BIN Fusion Research Team
Chonbuk National University
Dukjin, Jeonju 561-756
Korea

Suk Young Kim, PhD
School of Materials Science & Engineering
Yeungnam University
Gyeongsan, Gyeongbuk 712-749
Korea

Jong Tae Ko, MS
BK-21 Polymer BIN Fusion Research Team
Chonbuk National University
Dukjin, Jeonju 561-756
Korea

Doo Sung Lee, PhD
Polymer Science and Engineering
SungKyunKwan University
Jangan-ku, Suwon 440-746
Korea

Hai Bang Lee, PhD
Medical Science Division
Korea Research Institute of Chemical Technology
P.O. Box 107, Yuseong, Daejeon 305-600
Korea

Jin Ho Lee, PhD
Department of Advanced Materials
Hannam University
Daedeok-gu, Daejeon 306-791
Korea

Sang Jin Lee, PhD
Wake Forest Institute for Regenerative Medicine
Wake Forest University Health Sciences
Medical Center Boulevard, Winston-Salem
NC 27157
USA

Seung Jae Lee, MS
Laboratory of Tissue Engineering
Korea Institute of Radiological and Medical Sciences
Nowon-ku, Seoul 139-706
Korea

Seung Jin Lee, PhD
Department of Pharmacy
College of Pharmacy
Ewha Womans University
Seodaemun-gu, Seoul 120-750
Korea

Patrícia B. Malafaya, MS
3B's Research Group — Biomaterials, Biodegradables and
    Biomimetics
University of Minho, Campus of Gualtar
Braga 4710-057
Portugal

Byung-Hyun Min, MD, PhD
Department of Orthopedic Surgery
Ajou University Hospital
Paldalgu, Suwon 442-721
Korea

Nuno M. Neves, MS
3B's Research Group — Biomaterials, Biodegradables and
    Biomimetics
University of Minho, Campus of Gualtar
Braga 4710-057
Portugal

Se Heang Oh, PhD
Department of Advanced Materials
Hannam University
Daedeok-gu, Daejeon 306-791
Korea

Jung Su Park, MS
BK-21 Polymer BIN Fusion Research Team
Chonbuk National University
Dukjin, Jeonju 561-756
Korea

Jung Keug Park, PhD
Department of Chemical and Biochemical Engineering
Dongguk University
Choong-gu, Seoul 100-715
Korea

Ki Dong Park, PhD
Department of Molecular Science & Technology
Ajou University
Yeongtong, Suwon 443-749
Korea

Tae Gwan Park, PhD
Department of Biological Sciences
Korea Advanced Institute of Science and Technology
Yuseong, Daejon 305-701
Korea

Rui L. Reis, PhD
3B's Research Group — Biomaterials, Biodegradables and
    Biomimetics
University of Minho, Campus of Gualtar
Braga 4710-057
Portugal

Paula Sol, MS
3B's Research Group — Biomaterials, Biodegradables and
    Biomimetics
University of Minho, Campus of Gualtar
Braga 4710-057
Portugal

Youngsook Son, PhD
Laboratory of Tissue Engineering
College of Life Science
Kyung Hee University
Giheung-gu, Seoul 446-701
Korea

Soo Chang Song, PhD
Division of Life Science
Korea Institute of Science and Technology
Seongbuk, Seoul 130-650
Korea

Rui A. Sousa, MS
3B's Research Group — Biomaterials, Biodegradables and
    Biomimetics
University of Minho, Campus of Gualtar
Braga 4710-057
Portugal

Tao Xu, MS
Wake Forest Institute for Regenerative Medicine
Wake Forest University Health Sciences
Medical Center Boulevard, Winston-Salem
NC 27157
USA

James J. Yoo, MD, PhD
Wake Forest Institute for Regenerative Medicine
Wake Forest University Health Sciences
Medical Center Boulevard, Winston-Salem
NC 27157
USA

Chang Kook You, MS
School of Materials Science & Engineering
Yeungnam University
Gyeongsan, Gyeongbuk 712-749
Korea

So Hee Yun, MS
Laboratory of Biomaterials
R&D Institute, MCTT
Nowon-ku, Seoul 139-706
Korea

Introduction

*Gilson Khang,
Moon Suk Kim and
Hai Bang Lee*

It has been recognised that tissue engineering offers an alternative technique to tissue transplantation for diseased or malfunctioned organs. Millions of patients suffer from end-stage organ failure or tissue loss each year. In the United States alone, at least eight million surgical operations are carried out annually, requiring a total national healthcare cost exceeding US$400 billion per year [1, 2]. In the case of cardiovascular disease, approximately 500 000 coronary artery bypass surgeries are conducted each year in the United States [3]. Autologous and allogenic natural tissues, i.e. saphenous vein or internal mammary artery, are generally used for coronary artery replacement. The results have been quite favourable for these procedures, with patency rates generally ranging from 50% to 70%.

Failure in these procedures may be caused by intimal thickening, due in large part to the adaptation of the vessel in response to increased pressure and wall shear stress, compression, adequate graft diameter, and disjunction at the anastomosis. Successful treatment may also be limited by the poor performance of synthetic materials, such as polyethylene terephthalate (PET, Dacron®) and expanded polytetrafluoroethylene (ePTFE, Gore-Tex®), used for tissue replacement due to plagueing problems [4]. Despite improved patient outcomes, many of these materials possess serious problems including unpredictable outcomes, fibrous capsule contraction, allergic reactions, suboptimum mechanical properties, distortion, migration, and long-term resorption.

In order to avoid the shortage of donor organs and the abovementioned problems caused by the poor biocompatibility of biomaterials, a new hybridised method combining cells and biomaterials has been introduced: tissue engineering [5]. To reconstruct a new tissue by tissue engineering, triad components are needed. These include (1) cells which are harvested and dissociated from the donor tissue, including nerve, liver, pancreas, cartilage, and bone as well as embryonic/adult stem or precursor cell; (2) biomaterials as scaffold substrates in which cells are attached and cultured, resulting in implantation at the desired site of the functioning tissue; and (3) growth factors which promote and/or prevent cell adhesion, proliferation, migration, and differentiation by upregulating or downregulating the synthesis of protein, growth factors, and receptors (Fig. 1).

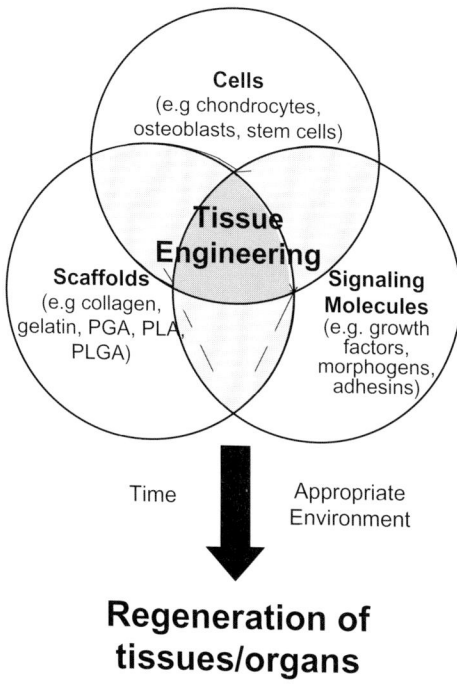

Regeneration of tissues/organs

Fig. 1 Tissue engineering triad. Consisting of three key elements (i.e. cells, biomaterials, and signalling molecules), it regenerates tissue-engineered neo-organs.

Importance of Scaffold Matrices in Tissue Engineering

Scaffolds play a critical role in tissue engineering. The function of scaffolds is to direct the growth of cells either seeded within the porous structure of the scaffold or migrating from surrounding tissue. The majority of mammalian cell types are anchorage-dependent, meaning they will die if an adhesion substrate is not provided. Scaffold matrices can be used to achieve cell delivery with high loading and efficiency to specific sites. Therefore, the scaffold must provide a suitable substrate for cell attachment, cell proliferation, differentiated function, and cell migration.

The prerequisite physicochemical properties of scaffolds are many: to support and deliver cells; induce, differentiate, and channel tissue growth; target cell-adhesion substrates; stimulate cellular response; provide a wound-healing barrier; be biocompatible and biodegradeable; possess relatively easy processability and malleability into desired shapes; be highly porous with a large surface/volume ratio; possess mechanical strength and dimensional stability; and have sterilisability, among others [2, 6]. Generally, three-dimensional porous scaffolds can be fabricated from natural and synthetic polymers, ceramics, metals, composite biomaterials, and cytokine release materials.

Natural polymers for scaffolds

Many naturally occurring scaffolds can be used as biomaterials for tissue engineering purposes. One example is the extracellular matrix (ECM), a very complex biomaterial controlling cell function that designs natural and synthetic scaffolds to mimic specific functions. Natural polymers include alginate, proteins, collagens (gelatin), fibrins, albumin, gluten, elastin, fibroin, hyarulonic acid, cellulose, starch, chitosan (chitin), scleroglucan, elsinan, pectin (pectinic acid), galactan, curdlan, gellan, levan, emulsan, dextran, pullulan, heparin, silk, chondroitin 6-sulfate, polyhydroxyalkanoates, etc. Much of the interest in these natural polymers comes from their biocompatibility, relative abundance and commercial availability, and ease of processing [8].

Synthetic polymers for scaffolds

Natural polymers are typically in short supply because they are expensive, suffer from batch-to-batch variation, and are susceptible to cross-contamination from unknown viruses or unwanted diseases. On the contrary, synthetic polymeric biomaterials have easily controlled physicochemical properties and quality, and have no immunogenicity. They can also be processed with various techniques and consistently supplied in large quantities. In order to adjust the physical and mechanical properties of tissue-engineered scaffolds at a desired place in the human body, the molecular structure and molecular weight are easily adjusted during the synthetic process.

Synthetic polymers are largely divided into two categories: biodegradeable and nonbiodegradeable. Some nonbiodegradeable polymers include polyvinylalcohol (PVA), polyhydroxyethymethacrylate (PHEMA), and poly(N-isopropylacrylamide) (PNIPAAm). Some synthetic biodegradeable polymers are the family of poly(α-hydroxy esters) such as polyglycolide (PGA), polylactide (PLA) and its copolymer poly(lactide-*co*-glycolide) (PLGA), polyphosphazene, polyanhydride, poly(propylene fumarate), polycyanoacrylate, poly(ε-caprolactone) (PCL), polydioxanone (PDO), and biodegradeable polyurethanes.

Of these two types of synthetic polymers, synthetic biodegradeable polymers are preferred for the application of tissue-engineered scaffolds because they minimise the chronic foreign body reaction and lead to the formation of completely natural tissue. That is to say, they can form a temporary scaffold for mechanical and biochemical support.

Bioceramics for scaffolds

Bioceramics are biomaterials that are produced by sintering or melting inorganic raw materials to create an amorphous or crystalline solid body, which can be used as an implant. Porous final products are mainly used for scaffolds. The components of ceramics are calcium, silica, phosphorus, magnesium, potassium, and sodium.

Bioceramics used for tissue engineering may be classified as nonresorbable (relatively inert), bioactive or surface active (semi-inert), and biodegradeable or resorbable (noninert). Alumina, zirconia, silicon nitride, and carbons are inert bioceramics; certain glass ceramics, such as dense hydroxyapatites [$9CaO \cdot Ca(OH)_2 \cdot 3P_2O_5$], are semi-inert (bioactive); and calcium phosphates, aluminium calcium phosphates, coralline, tricalcium phosphates ($3CaO \cdot P_2O_5$), zinc calcium phosphorus oxides, zinc sulfate calcium phosphates, ferric calcium phosphorus oxides, and calcium aluminates are resorbable ceramics. Of these bioceramics, synthetic apatite and calcium phosphate minerals, coral-derived apatite, bioactive glass, and demineralised bone particle (DBP) are widely used in hard tissue engineering.

Cytokine release system for scaffolds

Growth factors, a type of cytokine, are polypeptides that transmit signals to modulate cellular activities and tissue development such as cell patterning, motility, proliferation, aggregation, and gene expression. As in the development of tissue-engineered organs, the regeneration of functional tissue requires the maintenance of cell viability and differentiated function, encouragement of cell proliferation, modulation of the direction and speed of cell migration, and regulation of cellular adhesion. The easiest method for the delivery of growth factors is via injection near the site of cell differentiation and proliferation. However, this direct injection method incurs a relatively short half-life, a relatively high molecular weight and size, very low tissue penetration, and potential toxicity at the systemic level [2, 9].

One promising way to improve the efficacy of this technique is the locally controlled release of bioactive molecules for the desired release period by the impregnation into a scaffold. Through impregnation into a scaffold carrier, protein structure and biological activity can be stabilised to a certain extent, resulting in a prolonged release time at the local site. The duration of cytokine release from a scaffold is controlled by the types of biomaterials used, the loading amount of cytokine, the formulation factors, and the fabrication process. The cytokine release system may be designed for a variety of geometries and configurations, such as scaffold, tube, nose, microsphere, injectable forms, and fibre [10].

Fabrication and Characterisation for Scaffolds

Fabrication methods of scaffolds

Engineered scaffolds may enhance the functionalities of cell and tissue to support the adhesion and growth of a large number of cells by providing a large surface area and pore structure within a three-dimensional structure. Porosity provides adequate space, permits cell suspension, and penetrates the three-dimensional structure. These porous structures also promote ECM production, transport nutrients from nutrient media, and excrete waste products [6, 11]. Therefore, an adequate pore size as well as a uniformly distributed and interconnected pore structure are crucial to allow for easy distribution of cells throughout the scaffold structure. Scaffold structure is directly related to fabrication methods, many of which are listed in Table 1.

Table 1. The fabrication methods of scaffolds for tissue engineering.

Mechanism	Method	Remark
Leaching method	Solvent-casting/salt-leaching method	Chapter A
	Ice particle–leaching method	Chapter B
	Gas-foaming/salt-leaching method	Chapter C
	Gel-pressing method	Chapter D
Microsphere method	Biodegradeable microsphere	Chapter E
	Macroporous bead	Chapter F
	Particle-aggregated scaffold	Chapter G
Phase separation method	Freeze-drying method	Chapter H
	Thermally induced phase separation	Chapter I
	Centrifugation method	Chapter J
Injectable gel	Polyphosphazene gel	Chapter K
Acellular scaffold	Decellularisation process	Chapter L
Keratin scaffold	Self-assembled process	Chapter M
Fibre-spinning method	Nanofibre electrospinning process	Chapter N
	Microfibre wet spinning process	Chapter O
	Nonwoven PGA fibre	Chapter P
Printing and prototyping method	Inkjet printing process	Chapter Q
	Melt-based rapid prototyping	Chapter R
Functional scaffold	Growth factor release process	Chapter S
Ceramic scaffold	Sponge replication method	Chapter T
	Simple calcium phosphate coating method	Chapter U

This book introduces detailed protocols for 21 different types of manufacturing methods for scaffolds in Chapters A–U. The most common and commercialised one is the PGA nonwoven sheet (Albany International Research Co., Mansfield, MA, USA), which has a porosity of approximately 97% and a thickness of 1–5 mm. In order to dimensionally stabilise and provide the mechanical integrity, the fibre-bonding technology by heat and by PLGA/PLA solution spray coating methods has been developed [12] (Chapter P).

Porogen-leaching methods combine the polymerisation, solvent casting, gas foaming, or compression moulding of natural and synthetic scaffold biomaterials with the leaching of pore-generating particles (e.g. sodium chloride crystal, sodium tartrate, and sodium citrate) sieved using a molecular sieve [2, 11]. PLGA, PLA, collagen, poly(ortho ester), or small intestine submucosa (SIS)-impregnated PLGA scaffolds have successfully fabricated a biodegradeable sponge structure by this method with more than 93% porosity and a desired pore size of 1000 μm. Using the solvent-casting/particulate-leaching method, complex geometries such as tube, nose, and specific organ types can be fabricated as nanocomposite hybrid scaffolds by means of conventional polymer processing techniques like calendaring, extrusion, and injection. Complex geometries can be fabricated from the porous film lamination [13]. The advantage of this method is easy control of porosity and geometry. However, the disadvantages are the loss of water-soluble biomolecules or cytokines during porogen leaching, the possibility of remaining porogen as salt that can harmfully affect cell culture, and the different geometry surfaces and cross-sections (Chapters A–D).

The gas-foaming method refers to the exposure of a solid scaffold matrix to a sudden expansion of CO_2 gas under high pressure, resulting in the formation of a sponge structure due to nucleation and expansion in the dissolved CO_2 scaffold matrix. PLGA scaffolds with more than 93% porosity and about 100 μm median pore size have been developed using this method [14]. The significant advantage is no loss of bioactive molecules in the scaffold matrix, given that there is no need for the leaching process and no residual organic solvent; whereas the disadvantage is the presence of skimming film layers on the scaffold surface, resulting in the further removal process of this skin layer (Chapter C).

The phase separation method is divided into freeze drying, freeze thawing, freeze immersion precipitation, and emulsion freeze drying [15]. Phase separation by freeze drying can be induced by a polymer solution with an appropriate concentration by rapid freezing. The used solvent is then removed by freeze drying, resulting in porous structure as a portion of the solvent. Collagen scaffolds with pores of 50–150 μm, collagen–glycosaminoglycan blend scaffolds with an average pore size of 90–120 μm, and chitosan scaffolds with a pore size of 1–250 μm have been developed; the sizes vary with the freezing condition.

In addition, scaffold structures of synthetic polymers such as PLA and PLGA have been successfully made using this method, with over 90% porosity and 15–250 μm size.

The freeze-thawing technique induces phase separation between a solvent and a hydrophilic monomer upon freezing, followed by the polymerisation of the hydrophilic monomer by means of UV irradiation and removal of the solvent by thawing. This leads to the formation of macroporous hydrogel. A similar method is freeze immersion precipitation. The polymer solution is first cooled before being immersed in a nonsolvent and then a vapourised solvent, leading to porous scaffold structure. The emulsion freeze drying method is also useful for the fabrication of porous structure. In this case, a mixture of polymer solution and nonsolvent are thoroughly sonicated, quickly frozen in liquid nitrogen at −198°C, and then freeze-dried, resulting in sponge structure. The advantage of these phase separation techniques is the loading of hydrophilic or hydrophobic bioactive molecules, but the disadvantages are a relatively small pore size and difficulty in controlling the precise pore structure (Chapters H–J).

Injectable gel scaffolds have also been reported [11]. Injectable, gel-forming scaffolds provide several advantages: they can fill any shape of defect due to flowable materials, load various types of bioactive molecules and cells by simple mixing, do not contain residual solvents that may be present in a performed scaffold, and do not require surgical procedure for placement. Typical examples are thermosensitive gels such as Pluronics and polyethylene glycol (PEG)–PLGA–PEG triblock copolymer; pH-sensitive gels such as chitosan and its derivatives; ionically cross-linked gels such as alginate; fibrin gel; hyaluronan gel; etc. In the near future, multifunctional and tissue-specific gels, very fast sol–gel transition, and injectable scaffold materials that are fully degradeable for the desired period are expected to be available (Chapter K).

Nanoelectrospinning of PGA, PLA, PLGA, PCL copolymers, collagen, elastin, etc. has been extensively developed. For example, electrostatic processing can consistently produce PGA fibre diameters at or below 1 μm. By controlling the pick-up of these fibres, the orientation and mechanical properties can be tailored to a specific need of the injured site. Collagen electrospinning has also been performed, utilising type I collagen dissolved in HFP with 0.083 g/mL concentration. The optimally electrospun type I collagen nonwoven fabric appeared with an average diameter of 100 ± 40 nm, resulting in biomimicking of fibrous scaffolds (Chapters N–P).

Moreover, newly hybridised fabrication techniques, such as organic/inorganic and synthetic/natural at the nano-sized level, are being continuously developed for the application of tissue-engineered scaffolds.

Physicochemical characterisation of scaffolds

For the successful achievement of three-dimensional scaffolds, several characterisation criteria are required. They can be divided into four categories: (1) morphology (e.g. porosity, pore size, surface area); (2) mechanical properties (e.g. compressive and tensile strength); (3) bulk properties (e.g. degradation and its relevant mechanical properties); and (4) surface properties (e.g. surface energy, chemistry, charge).

Porosity is defined as the fraction (i.e. percentage) of the total volume occupied by voids. The most widely used methods for measuring porosity are mercury porosimetry, scanning electron microscopy, and confocal laser microscopy.

Mechanical properties are extremely important when designing tissue-engineered products. To determine the mechanical properties of a porous structure, conventional testing instruments may be used. Mechanical tests can be divided into creep tests, stress–relaxation tests, stress–strain tests, and dynamic mechanical tests. These test methods are similar to those of conventional biomaterials.

The rate of degradation of manufactured scaffolds is one of the most important factors in designing tissue-engineered products. Ideally, the scaffold construct provides mechanical and biochemical support until the entire tissue regenerates without any change, and then it completely biodegrades at a rate consistent with tissue generation. Immersion studies are commonly conducted to track the degradation of biodegradeable matrices. So, the changes in weight loss and molecular weight can be evaluated by the chemical balance, scanning electron microscopy, and gel permeation chromatography. From these results, the mechanism of biodegradation can be determined.

It is generally recognised that the adhesion and proliferation of different types of cells on polymeric materials depend largely on surface characteristics such as wettability (hydrophilicity/hydrophobicity of surface free energy), chemistry, charge, roughness, and rigidity. In particular, three-dimensional applications of tissue engineering are more important for cell migration, cell proliferation, DNA/RNA synthesis, and phenotype presentation on the scaffold materials. Surface chemistry and charge can be analysed by electron scanning chemical analysis and streaming potential, respectively. The wettability of the scaffold surface can be measured by the contact angle with static and dynamic methods.

References

[1] Langer R, Vacanti J. *Science* **260**: 920–926, 1993.
[2] Khang G, Lee SJ, Kim MS, Lee HB. Biomaterials: tissue engineering and scaffold, in Webster J (ed.), *Encyclopedia of Medical Devices and Instrumentation*, 2nd ed., Wiley Press, New York, 366–383, 2006.
[3] Mann BK, West JL. *Anat Rec* **263**: 367–371, 2001.
[4] Lee HB, Khang G, Lee JH. Polymeric biomaterials, in Park JB, Bronzino JD (eds.), *Biomaterials: Principles and Applications*, CRC Press, Boca Raton, FL, 2003.
[5] Petit-Zeman S. *Nat Biotechnol* **19**: 201–206, 2001.
[6] Chaignaud BE, Langer R, Vacanti JP. The history of tissue engineering using synthetic biodegradable polymer scaffolds and cells, in Atala A, Mooney DJ (eds.), *Synthetic Biodegradable Polymer Scaffolds*, Birkhauser, Boston, MA, 1996.
[7] Wong WH, Mooney DJ. Synthesis of properties of biodegradable polymers used as synthetic matrices for tissue engineering, in Atala A, Mooney DJ (eds.), *Synthetic Biodegradable Polymer Scaffolds*, Birkhauser, Boston, MA, 1996.
[8] Baldwin SP, Saltzman WM. *Adv Drug Deliv Rev* **33**: 71–86, 1998.
[9] Khang G, Kim MS, Min BH, Lee I, Rhee JM, Lee HB. *Tissue Eng Reg Med* **3**: 376–395, 2006.
[10] Seal BL, Otero TC, Panitch A. *Mater Sci Eng R* **34**: 147–230, 2001.

[11] Thompson RC, Wake MC, Yasemski MJ, Mikos AG. *Adv Polym Sci* **122**: 245–274, 1995.

[12] Khang G, Shin P, Kim I, Lee B, Lee SJ, Lee YM, Lee HB, Lee I. *Macromol Res* **10**: 158–167, 2002.

[13] Leibmann-Vinson A, Hemperly JJ, Guarino RD, Spargo CA, Heidaran MA. Bioactive extracellular matrices: biological and biochemical evaluation, in Lewandrowski KU, Wise DL, Trantolo DJ, Gresser JD, Yasemski MJ, Altobeli DE (eds.), *Tissue Engineering and Biodegradable Equivalents: Scientific and Clinical Applications*, Marcel Dekker, New York, 2002.

[14] Khang G, Jeon JH, Cho JC, Lee HB. *Polymer(Korea)* **23**: 471–177, 1999.

A

Protocol for Solvent-Casting/ Salt-Leaching Method

Gilson Khang, Soon Hee Kim and Hai Bang Lee

A.1 Concept

- In tissue engineering, absorbable polymer (PLLA, PLGA, etc.) scaffolds are used to support cells until they are replaced by the body's own ECM.
- An ideal scaffold should be biocompatible, biodegradeable, and highly porous with interconnected pores.
- Porous three-dimensional temporary scaffolds play an important role in manipulating cell function in terms of the formation of the new organ or tissue.
- To prepare three-dimensional biodegradeable porous scaffolds, a method that incorporates the use of salt particles as the porogen material is described below.
- The porogen leaching method provides easy control of the pore structure. The pore structure, porosity, and pore size can be easily controlled by regulating the amount and size of salt.
- This method involves casting a mixture of polymer solution (polymer/chloroform or polymer/methylene chloride) and porogen in a mould, and then leaching out the porogen with water to generate the pores and freeze-drying the mixture.
- Water-soluble particulates, such as salts and carbohydrates, are used as the porogen materials.

A.2 Procedure

- Salt particulates are prepared by sieving. The sizes of the salt particulates are controlled by the desireable sieving [Fig. A.1(a)].
- Polymer solutions are prepared by dissolving different amounts and types of polymers in solvent (e.g. methylene chloride or chloroform) [Fig. A.1(b)].
- Sieved salt particulates are added to the polymer solution, and the dispersion is gently vortexed [Fig. A.1(c)].
- The solution is poured into the designed silicon mould [Fig. A.1(d)].
- Subsequently, the mould with dispersion is pressed (60 kgf/cm^2) by pressure apparatus [Fig. A.1(e)].

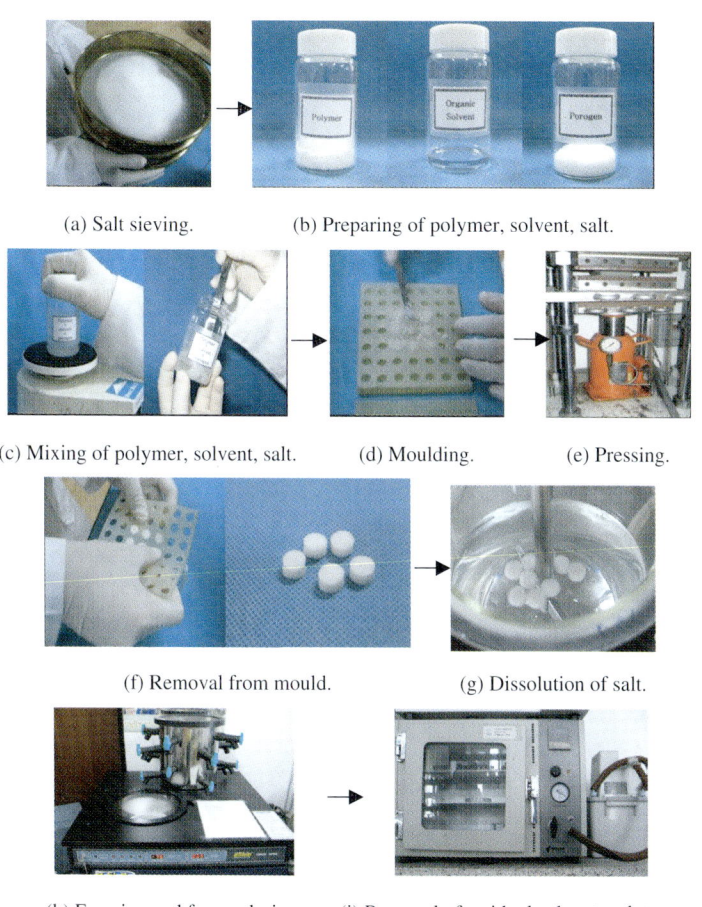

(a) Salt sieving. (b) Preparing of polymer, solvent, salt.

(c) Mixing of polymer, solvent, salt. (d) Moulding. (e) Pressing.

(f) Removal from mould. (g) Dissolution of salt.

(h) Freezing and freeze-drying. (i) Removal of residual solvent and storage.

Fig. A.1 The preparation processing of scaffold by salt-leaching method.

- The formed samples are taken out of the mould [Fig. A.1(f)].
- Samples are dissolved for a desireable time (48 h) in deionised water [Fig. A.1(g)].
- Salt-removed samples are freeze-dried for a desireable time (about 48 h) at low temperature (8 mTorr, −55°C) [Fig. A.1(h)].
- The scaffolds are dried in a vacuum oven at 25°C for 1 week to remove the residual solvent. Scaffolds are kept under vacuum until use [Fig. A.1(i)].

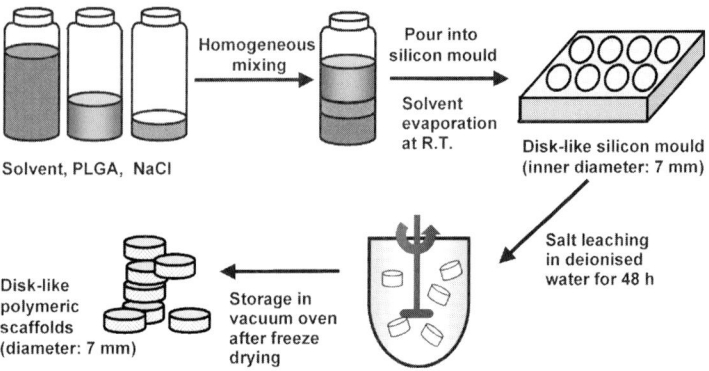

Fig. A.2 Schematic diagram of the salt particle-leaching method to fabricate PLGA scaffold.

A.3 Requirements

1. Polymer
2. Sieve
3. Solvent
4. Vortexer
5. Spatula
6. Deionised water
7. Mould
8. Freeze dryer

A.4 Characterisations

- Using the demanded sieve, the sizes of the salt particulates are controlled. The shape of the salt is measured through photomicrographs.
- The salt particulates are almost square-shaped (Fig. A.4).
- The salt-leaching method provides high porosity up to 97%, and median pore diameters up to 140 μm are prepared.
- The porosity and pore size can be independently controlled by varying the amount and size of the salt particles, respectively.
- The porosity and pore size, according to the size of salt particles, are analysed by the mercury intrusion method (Table A.1).

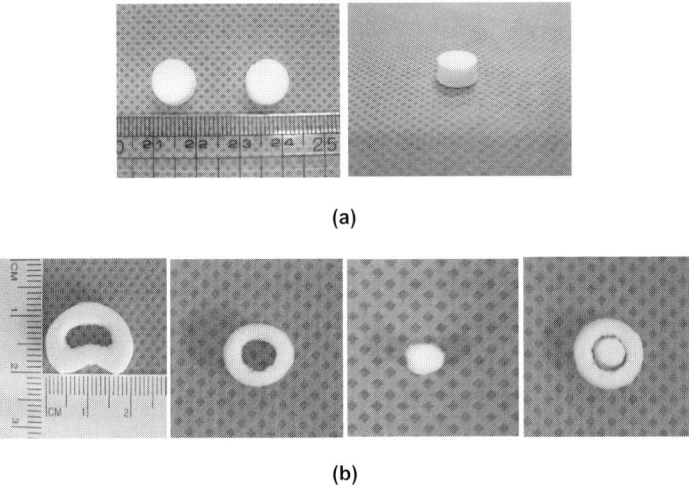

(a)

(b)

Fig. A.3 (a) The shape of completed scaffold, and (b) various images of porous 3D scaffolds fabricated by salt-leaching method.

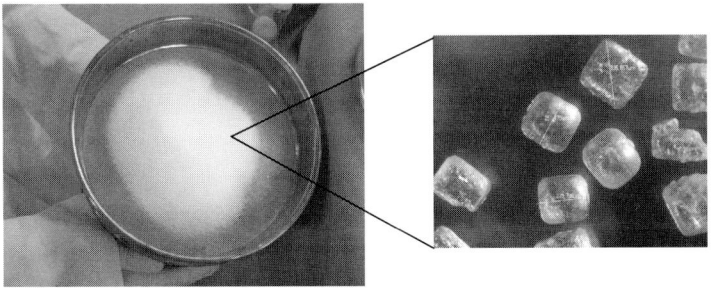

Fig. A.4 Image of sieved salt microparticle.

- The theoretical porosity can be calculated from the weight fraction of salt particulates to polymer and from the densities of polymer and salt.
- The pore structures, according to the mixing of desireable materials (e.g. ipriflavone, DBP), are analysed by the mercury intrusion method (Fig. A.5).
- The porosity and pore size are not heavily affected by mixed material.

Table A.1 Properties of fabricated porous PLGA scaffolds by means of solvent casting/salt leaching.

Type of polymer	Size of NaCl (μm)	PLGA concentration (w/v%)	Volume of PLGA to NaCl (w/w%)	Porosity (%)	Median pore diameter (μm)
PLA	180~250	20	90	94.1	92.1
PLA	250~355	20	90	92.5	109.3
PLGA	180~250	20	90	97.3	120.8
PLGA	250~355	20	90	96.5	133.7

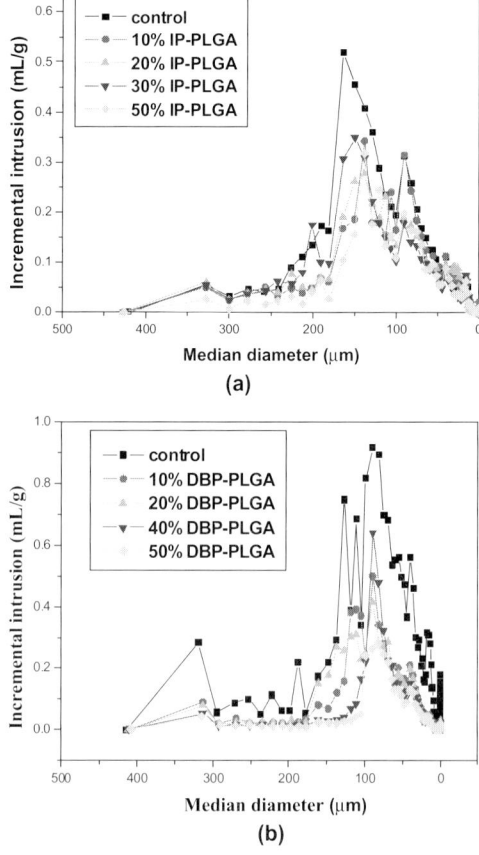

Fig. A.5 Pore size distribution of scaffolds mixing desireable materials by means of solvent casting/salt-leaching.

Table A.2 Amount of residual NaCl with variation of salt extraction in water.

Time (h)	Residual NaCl
6	1.92 mg/mL
12	ND*
24	ND
48	ND

*ND: no detection.

- The residual salt amount in completed scaffold is not detected after 12 h of salt leaching (Table A.2).
- The salt amount remaining in scaffold, according to the salt-leaching time, is analysed using ion chromatography.
- The pore structures are observed by SEM.
- The pore structures of the three-dimensional scaffolds can be regulated by controlling the properties of the salt particulates.
- Salt-leached scaffolds show a typical square-shaped (not collapsed) pore structure, with large pore sizes corresponding to the sieved salt particle size and smaller interstitial pores between the salt particles [Fig. A.6(a)].
- The cross-section of the PLGA scaffold seems to have an interconnected network structure and comparably regular pores [Fig. A.6(b)].
- The pore shapes are almost the same as those of the salt particulates (Fig. A.7).

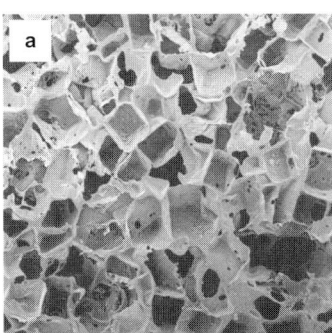

Fig. A.6 SEM pictures of PLGA scaffold fabricated by salt-particle leaching method (×50). (a) Surface; (b) cross-section.

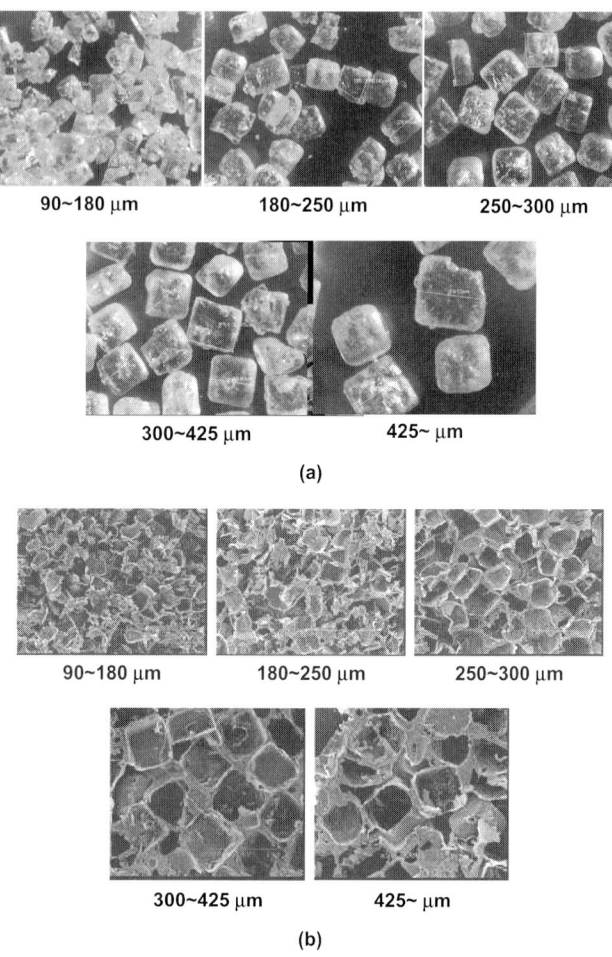

Fig. A.7 (a) Various sizes of salt particles, and (b) surface of PLGA scaffold fabricated by salt-particle leaching method according to pore size (×50).

A.5 Cautions

- Scaffold-impregnated soluble protein cannot be manufactured using this method.

(*Continued*)

(*Continued*)

- To form open and interconnected pores, the proper salt size and amount must be selected.
- The salt-leaching period must be properly applied in the manufacture process.
- Residual solvent must be removed in scaffolds.
- The surface of scaffold may be stopped due to high-pressure solvent; therefore, the surface of scaffold must be truncated.
- Because complete removal of salt from the centre of the scaffold is difficult, this method may not apply to thicker scaffolds greater than 2 mm.

B

Protocol for Ice Particle–Leaching Method

*Moon Suk Kim,
Gilson Khang and
Hai Bang Lee*

B.1 Concept

- To prepare porous 3D scaffolds for tissue engineering, the porogen-leaching method provides easy control of pore structure and has been widely utilised.
- Recently, several different water-soluble particles, including salts and carbohydrates, have been used as the porogen material.
- Scaffold manufacturing usually involves (1) dissolving the polymer in organic solvent, (2) incorporating porogens, and (3) leaching porogens.
- The pore structure, porosity, pore size, and pore morphology can be easily manipulated by controlling the properties of the porogen, and the process is reproducible.
- The prepared porous 3D scaffolds may support cell growth both *in vitro* and *in vivo.*
- Despite these advantages, the problem of residual porogen used to prepare 3D scaffolds remains.
- Therefore, the conventional method of porogen leaching by washing with water is replaced by freeze-drying, facilitating the removal of the porogen and making removal more complete.
- The method of porogen leaching by using ice particulates as the porogen material can be employed to fabricate porous 3D scaffolds for tissue engineering.
- Using ice particulates as the porogen material, scaffolds are prepared by mixing a polymer solution in a solvent with ice particulates, freezing the mixture in liquid nitrogen, and freeze-drying.
- This method can be applied to polymers that are soluble in a solvent such as chloroform or methylene chloride.
- Biodegradeable polymers of PLLA and PLGA can be utilised for this method.
- Sieved ice particles are dispersed in a polymer/chloroform solution that is used to fabricate porous 3D scaffolds.
- The ice particles are eventually leached out by selective dissolution in water or by freeze-drying to produce a porous 3D scaffold.

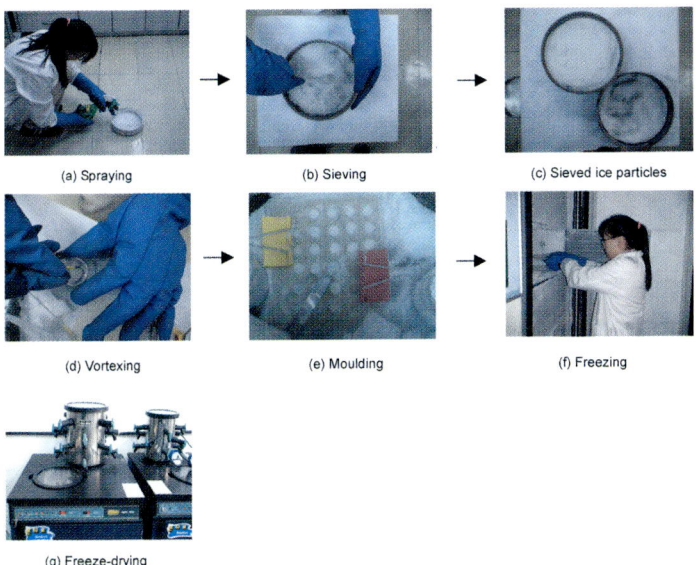

(a) Spraying

(b) Sieving

(c) Sieved ice particles

(d) Vortexing

(e) Moulding

(f) Freezing

(g) Freeze-drying

Fig. B.1 Schematic diagram of the preparation processing of scaffolds by the ice particle–leaching method.

B.2 Procedure

- Ice particulates are prepared by spraying deionised water into liquid nitrogen [Fig. B.1(a)].
- The sizes of the ice particulates are controlled by the desireable sieving [Figs. B.1(b) and B.1(c)].
- Polymer solutions of various concentrations are prepared by dissolving different amounts of polymer in solvent (e.g. methylene chloride or chloroform) and cooling the solution to −20°C.
- Ice particulates are added to the precooled polymer solution.
- The dispersion is gently vortexed [Fig. B.1(d)].
- It is then poured into a precooled designed mould [Fig. B.1(e)].
- Subsequently, the mould with dispersion is frozen by placing in low temperature [Fig. B.1(f)].
- The mould with dispersion is freeze-dried for a desireable time under low temperature [Fig. B.1(g)].
- (Often, further drying at elevated temperatures is required to completely remove the solvent after freeze-drying.)

B.3 Requirements

1. Polymer
2. Liquid nitrogen
3. Deionised water
4. Sieve
5. Solvent
6. Vortexer
7. Mould
8. Freeze dryer

B.4 Characterisations

- The sizes of the ice particulates are controlled by the desireable sieve and measured from their photomicrographs.
- The ice particulates are almost spherical (Fig. B.2).
- Their diameters are measured from their photomicrographs by hypothesising that all of their shapes are spherical.
- Various scaffold forms can be easily manipulated by a designed mould using the ice particle–leaching method (Fig. B.3)..
- The prepared scaffolds are physically stable and manageable.
- Highly porous 3D scaffolds with porosities up to 99% and median pore diameters up to 400 μm have been prepared using the ice-particle leaching method.

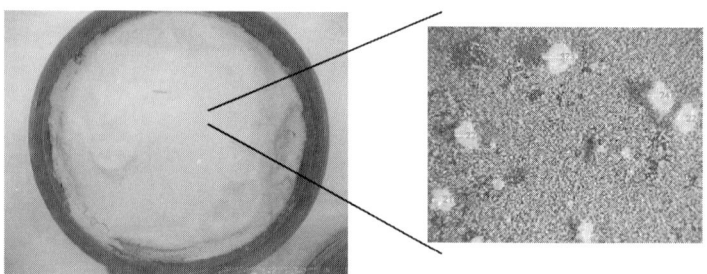

Fig. B.2 Image of sieved ice microparticle.

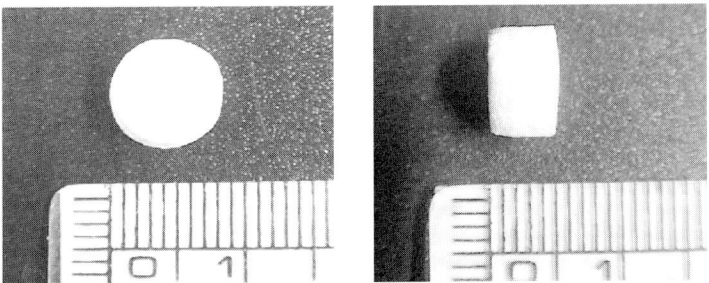

Fig. B.3 Image of porous 3D scaffolds fabricated by the ice particle–leaching method.

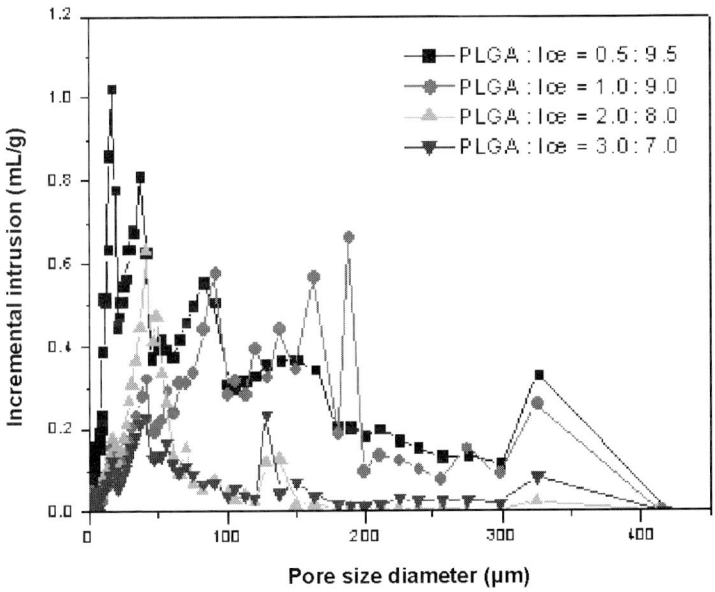

Fig. B.4 Pore sizes of PLGA scaffolds fabricated by the ice particle–leaching method with different ratios of polymer and ice.

- Their pore structures are analysed by the mercury intrusion method (Fig. B.4).
- The median pore size, porosity, and surface area of the scaffolds are determined by mercury intrusion porosimetry and are summarized in Table B.1.

Table B.1 Preparation of PLGA scaffolds with different ratios of polymer and ice.

Sample	Porosity (%)	Total pore area (m^2/g)	Median pore diameter (µm)
PLGA:Ice = 0.5:9.5	93.6	1164.0	27.9
PLGA:Ice = 1.0:9.0	99.7	575.1	51.1
PLGA:Ice = 2.0:8.0	99.1	196.1	37.1
PLGA:Ice = 3.0:7.0	89.1	127.9	36.0

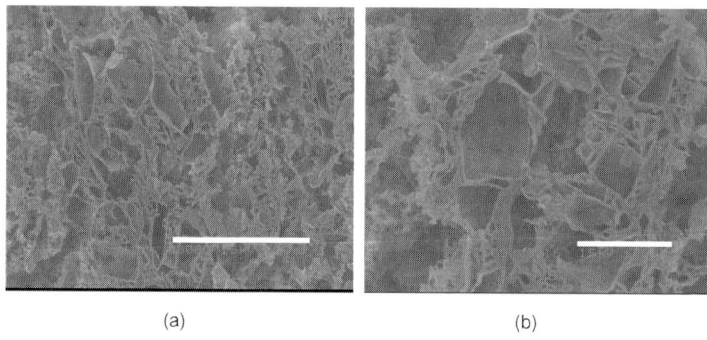

(a)   (b)

Fig. B.5 SEM pictures of PLGA scaffold fabricated by the ice particle–leaching method. Magnification is (a) 100 and (b) 300, and the scale bar represents (a) 500 µm and (b) 100 µm.

- The mean pore diameter is ~50 µm.
- The theoretical porosity can be calculated from the weight fraction of ice particulates to polymer and the densities of polymer and ice.
- Their pore structures are observed by SEM (Fig. B.5).
- The pore structures of the 3D scaffolds can be manipulated by controlling the properties of the ice particulates and the polymer concentration.
- The cross-section of PLGA scaffold seems to have an interconnected network structure and comparably regular pores.
- The pore shapes are almost the same as those of the ice particulates.
- The scaffolds become more interconnected as the weight fraction of the ice particulates increases.

B.5 Cautions

- This process must be conducted in a cold condition.
- During the freezing in liquid nitrogen, phase separation may have occurred in the polymer solution, resulting in a deformed porous structure after freeze-drying.
- The ice particulates reformed during freezing are influenced by several processing variables such as the temperature of freezing, thus making it difficult to precisely control the pore structure, including the pore size distribution and the surface area of the scaffold.

C

Protocol for Gas-Foaming/ Salt-Leaching Method

Tae Gwan Park

C.1 Concept

- The particulate leaching technique has been widely utilised to fabricate three-dimensional porous scaffolds for tissue engineering. As a porogen particle, various salt and other water-soluble particles have been utilised for this technique.
- However, the biodegradeable scaffolds prepared by the particulate-leaching method often exhibit a dense surface skin layer, which hampers *in vitro* cell seeding into the scaffolds and tissue ingrowth after *in vivo* implantation.
- Additionally, poor interconnectivities between macropores lower cell viability and result in nonuniform distribution of seeded cells throughout the matrix.
- Sodium bicarbonate salt or ammonium bicarbonate salt with acidic excipients such as citric acid has been used for effervescent gas-evolving oral tablets, due to its carbon dioxide–evolving property upon contact in acidic aqueous solution. Thus, various alkalinising analgesic oral tablets are commercially available.
- In particular, ammonium bicarbonate salt — upon contact in an acidic aqueous solution and/or incubated at elevated temperature — evolves gaseous ammonia and carbon dioxide by itself.
- The gas-foaming/salt-leaching method is based on the idea that sieved salt particles of ammonium bicarbonate dispersed within a polymer–solvent mixture can generate ammonia and carbon dioxide gases within solidifying matrices upon contact with hot water or aqueous acidic solution, thereby producing highly porous structures (Fig. C.1).
- These scaffolds show macropore structures over 200 µm with no visible surface skin layers, thus permitting sufficient cell seeding within the scaffolds.
- In addition, porosities can be controlled by the amount of ammonium bicarbonate incorporated into the polymer solution.
- It is possible to make various scaffolds with different geometries and sizes by a hand-shaping or moulding process because the polymer–salt mixture becomes a gel paste after partial evaporation of the solvent.

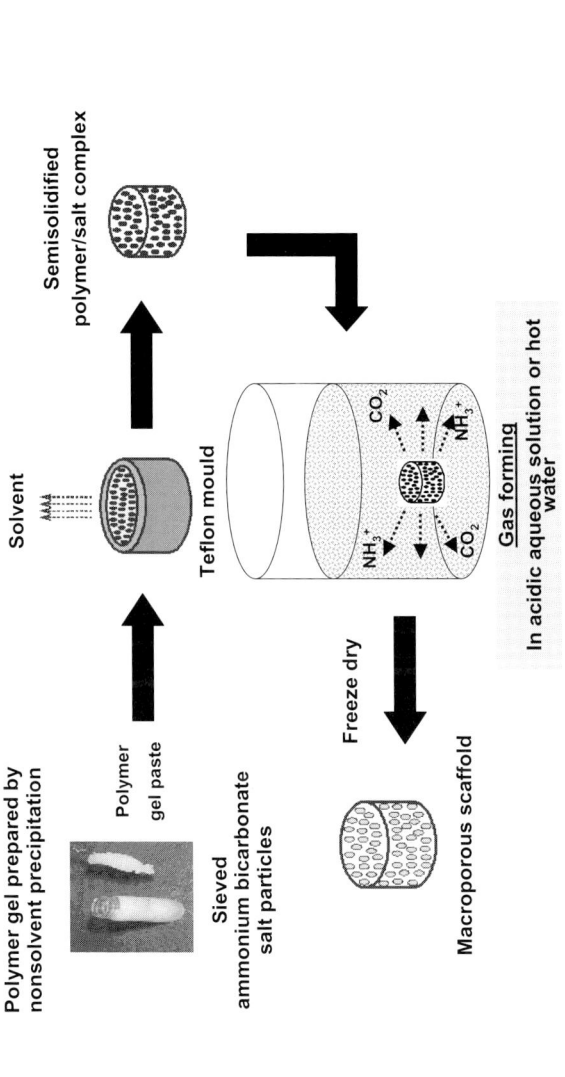

Fig. C.1 Schematic illustration of macroporous biodegradeable scaffolds fabricated by the gas-foaming/salt-leaching method.

C.2 Materials

1. Ammonium bicarbonate salt particles

   - Ammonium bicarbonate crystals are grinded into smaller particles.
   - These salt particles are separated with sieves into different size ranges (100–200 and 300–500 μm).
   - Sieved salt particles should be sealed with parafilm in a glass bottle to prevent moisture and particle aggregation.

2. Polymer solution

   - PLLA weight-average molecular weight of 300 000 g/mol (Polysciences, Inc., Warrington, PA).
   - PLGA Medisorb®, lactic/glycolic molar ratios of 75/25 (Mw 97 200), 65/35 (Mw 74 300), and 50/50 (Mw 54 200) (Alkermis, Cincinnati, OH).
   - Filtered organic solvent (such as chloroform) and sterilised glass scintillation vials are recommended to make a clean polymer solution.

3. Teflon moulds

   - Teflon plate having two different dimensions of 10 mm in diameter with 2-mm and 5-mm thickness.
   - Mould is kept clean before use.

4. Citric acid aqueous solution

   - Citric acid is dissolved in distilled deionised water to make supersaturated solution.
   - Citric acid aqueous solution is filtered with a 0.45-μm filter.

C.3 Gas Foaming/Salt Leaching in Acidic Aqueous Solution

- Completely dissolve 1 g of PLGA in 10 mL of chloroform to make 10% polymer solution.
- Add excess volume of cold ethanol to the polymer solution.
- Mix homogeneously. A gel-like slurry precipitates immediately in the solvent/nonsolvent mixture [see Caution 1(a)].

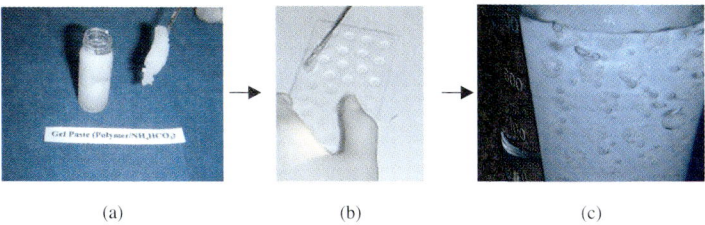

Fig. C.2 (a) Gel paste of polymer/salt (ammonium bicarbonate)/solvent (nonsolvent) mixture. (b) Photograph of the casting procedure using a disc-shaped Teflon mould. (c) Gas-foaming procedure using supersaturated citric acid solution.

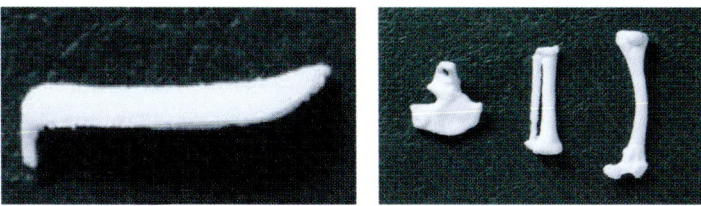

Fig. C.3 Photographs of nasal prosthesis and bone-shaped polymer scaffolds prepared by bone-shaped silicon moulds.

- Remove turbid solution and recover the gel slurry.
- Add 10 g of ammonium bicarbonate to the solution (see Cautions 2, 3, and 4).
- Mix homogeneously [a small volume (~2 mL) of chloroform can be added to the slurry as a plasticiser] to make a homogeneous gel paste mixture of polymer/salt [Fig. C.2(a)].
- Cast the paste mixture into a disc–shaped Teflon mould [Fig. C.2(b)] or manipulate to the desired shape (Fig. C.3; see Caution 5).
- Dry the gel paste mixture by partial evaporation of ethanol under atmospheric pressure for 1 h to obtain the semisolidified mixture.
- Detach a polymer/salt complex from the mould.
- Wet the semisolidified polymer/salt complex with cold ethanol.
- Immerse the matrix into supersaturated citric acid solution to effervescence from embedded salt particles [Fig. C.2(c)].
- After complete effervescence, wash the scaffolds with dH_2O three times (see Caution 6).

- Soak the scaffolds three times in $3dH_2O$ for 2 h to remove residual salt particles.
- Freeze-dry the scaffolds for 5 days.
- Store at −80°C with desiccant until use.

C.4 Gas Foaming/Salt Leaching in Hot Water

- Dissolve PLLA polymer in chloroform at a concentration of 8% (w/v).
- Add ammonium bicarbonate salt particulates to the PLLA solution.
- Mix thoroughly with a spatula to make a homogeneous gel paste mixture of polymer/salt/solvent [see Caution 1(b)].
- Cast the paste mixture into a disc-shaped Teflon mould.
- Dry the paste mixture to partially evaporate chloroform under atmospheric pressure for 2 h.
- Wet the semisolidified mixture with cold ethanol.
- Immerse the mixture in an excess amount of hot water (90°C) until no gas bubbles are generated (~5 min; see Caution 6).
- Soak the samples in cold water for 30 min to remove residual salt particles, and rinse off the scaffolds.
- Freeze-dry for 5 days, and store at −80°C with desiccant until use.

C.5 Cautions

1(a) Amorphous PLGA dissolved in chloroform does not become a gel paste even at high concentration. As an alternative method, PLGA dissolved in chloroform is precipitated in nonsolvent such as ethanol.

1(b) The gel formation of semicrystalline PLLA in chloroform is caused by the physical cross-linking of crystalline domains between PLLA chains under the condition of highly concentrated PLLA polymer solution.

2. The weight ratio of NH_4HCO_3 to polymer can be adjusted at 10:1 or 20:1.

(Continued)

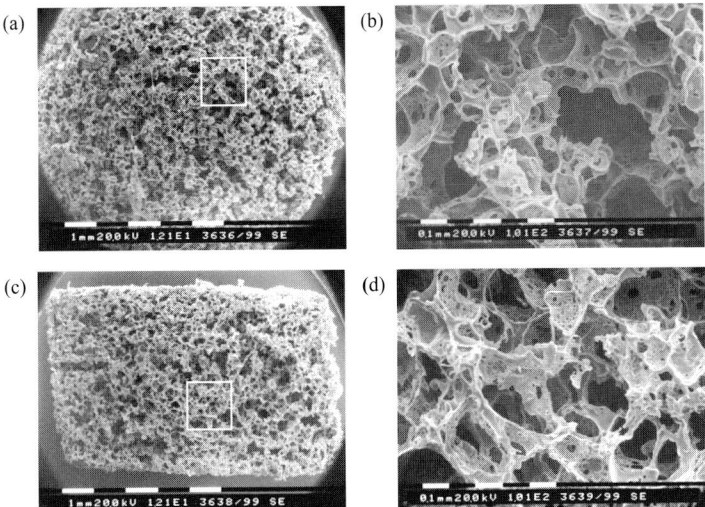

(a) (b)
(c) (d)

Fig. C.4 SEM images of thick PLGA scaffolds (thickness 5 mm). (a, c) Surface; (b, d) cross-section.

(Continued)

3. The salt particle size of ammonium bicarbonate salt can be controlled to 100–200 μm or 300–500 μm by using standard sieves.
4. The pore size and porosity can be controlled by the salt-particle–to–polymer-weight ratio and by the particle size.
5. Disc-shaped Teflon moulds having two different dimensions of 10 mm in diameter with 2 mm or 5 mm in thickness are used (Fig. C.4).
6. The thermal property of polymers in the scaffolds measured by DSC does not change significantly after effervescence.

D

Protocol for Gel-Pressing Method

Soo Hyun Kim

D.1 Concept

- The main focus of tissue engineering is the synthesis of artificial constructs or tissues based on vital cells or cell matrix.
- Biomaterials provide a three-dimensional structure to shape or guide tissue development.
- To successfully engineer functional tissues and organs, the biodegradeable scaffolds have to be designed to facilitate cell distribution and guide tissue regeneration in three dimensions.
- Several methods have been developed to create highly porous scaffolds, including the solvent-casting/particulate-leaching process, phase separation, gas foaming, and nanostructure scaffolds.
- The particulate-leaching process dissolves the polymer (PLLA or PLGA) in chloroform, and then casts it onto a dish filled with the porogen. After evaporation of the solvent, the polymer/salt composite is leached in water to remove the porogen. The process is easy to carry out.
- The pore size can be controlled by the size of the salt crystals, and the porosity by the salt/polymer ratio. However, certain critical variables such as pore shape, limited membrane thickness (3 mm), plastic operation, and interpore openings are not controllable.
- To overcome these shortcomings, a method to fabricate porous, biodegradeable scaffolds using a combined gel-pressing method and salt-leaching technique has been developed.
- This fabrication method can be successfully applied to a wide variety of polymer and salt formulations.

D.2 Procedure

- A polymer/salt composite is firstly prepared by dissolution process in a solvent [Fig. D.1(a)].
- The polymer is dissolved in a solvent and then mixed with salts [Fig. D.1(b)].
- The solvent is evaporated under air condition to form gels [Fig. D.1(c)].
- Polymer gels are pressed to fabricate a tubular or sheet-type scaffold [Figs. D.1(d)–D.1(f)].

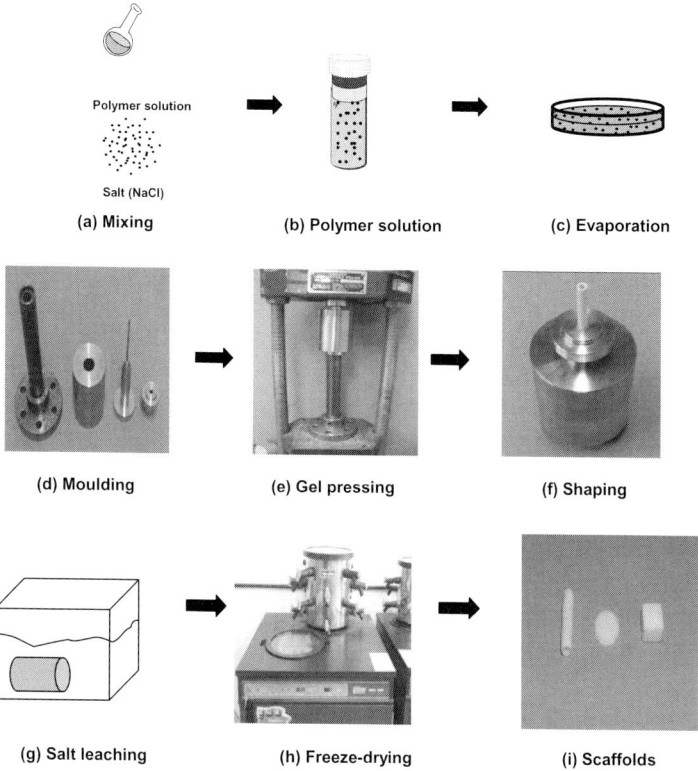

Fig. D.1 Fabrication of scaffolds by gel-pressing method.

- After evaporation of the solvent, the salt particles in the construct are leached out to make an open-pore structure [Fig. D.1(g)].
- The scaffolds are freeze-dried for a desired time under low temperature [Fig. D.1(h)].

D.3 Requirements

1. Polymer
2. Solvent
3. Salt

(*Continued*)

(*Continued*)

4. Freeze-dryer
5. Press
6. Mould
7. Deionised water

D.4 Characterisations

- The porosity, pore size, and macroscopic dimension of scaffolds are the most important factors associated with cell adhesion and proliferation.
- For tissue engineering, it is necessary to obtain a maximal supply of nutrition by diffusion from tissue culture media *in vitro* or through newly formed blood vessels *in vivo* by controlling the pore characteristics.
- The pore morphology of scaffolds prepared by the gel-pressing method is defined mainly by the size and shape of the salt crystals.
- The salt weight fraction is the most significant parameter affecting the porosity of the scaffold. An increase in salt weight fraction results in an increase in porosity.
- The pore size is affected by both the salt weight fraction and the size of the salt particles, and can be controlled by the salt size.
- The scaffolds by gel-pressing method have highly porous structures, high pore interconnectivities, no thickness limitations, good mechanical strength, and plastic operation as advantages.

D.5 Example

- We have developed a new technique for manufacturing a very elastic tubular or sheet-type scaffold by a gel-pressing method (Fig. D.1).

(*Continued*)

(Continued)

- To design a biodegradeable elastic scaffold (mechanical force–responsive scaffold, mechanoactive scaffold), a scaffold was fabricated with elastomeric PLCL (50:50).
- Ten grams of PLCL (a biodegradable polymer) was first dissolved in 50 mL of chloroform, and salt particles sieved to a certain size range (300–500 μm) were added (50%–90% wt/wt).
- Chloroform was evaporated under air condition to form gels.
- PLCL gels were pressed in a tubular or sheet-forming mould.
- The residual chloroform was evaporated for 48 h at room temperature and completely removed under vacuum for 24 h.
- The salts were leached out in distilled, deionised water with shaking for 3 days.
- The resultant scaffolds had highly porous and open-pore structures (Figs. D.2 and D.3).
- The mechanical properties of the PLCL scaffolds were highly elastic and flexible, which are very important factors

(Continued)

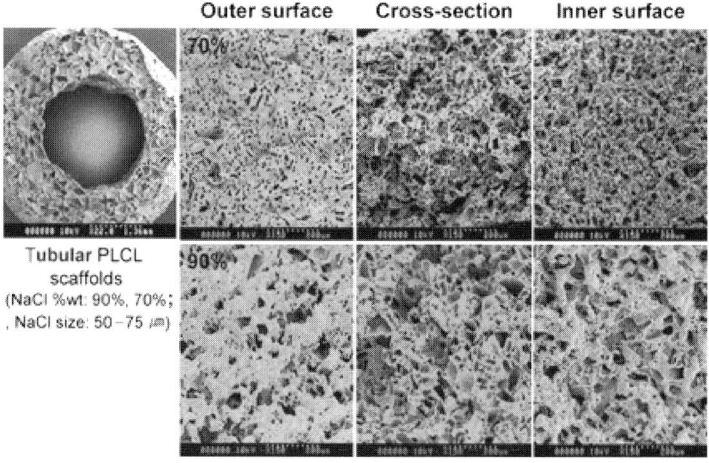

Fig. D.2 SEM images of PLCL tubular scaffolds by gel-pressing method.

(*Continued*)

for the fields of mechanoactive tissue engineering such as vascular graft tissue engineering or cartilage tissue engineering.

- The PLCL scaffolds had a very low tensile modulus and a very high elongation at break (Fig. D.4), and maintained a high recovery at applied strains of up to 500% (Fig. D.5).

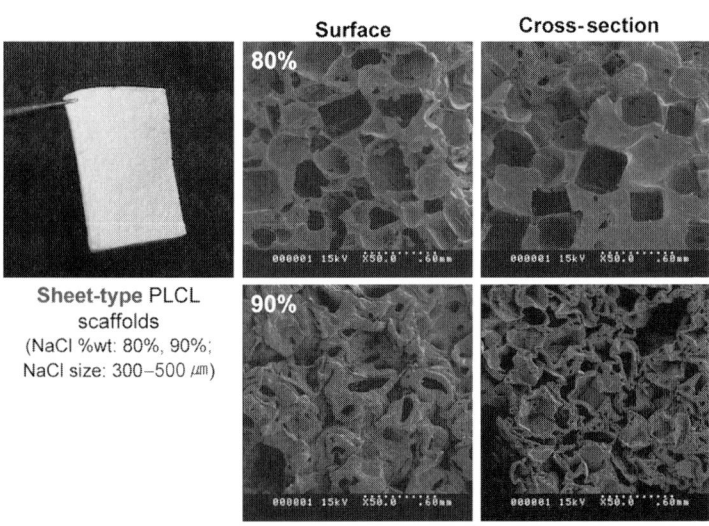

Fig. D.3 SEM images of PLCL sheet-type scaffolds by gel-pressing method.

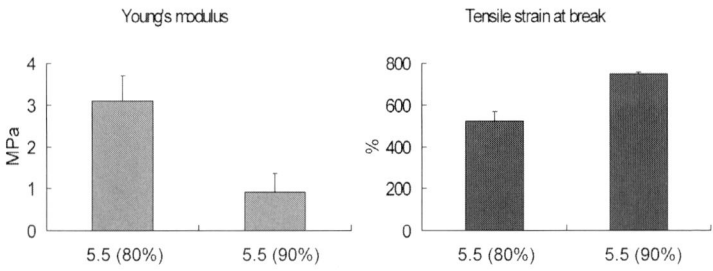

Fig. D.4 Mechanical properties of PLCL scaffolds by gel-pressing method.

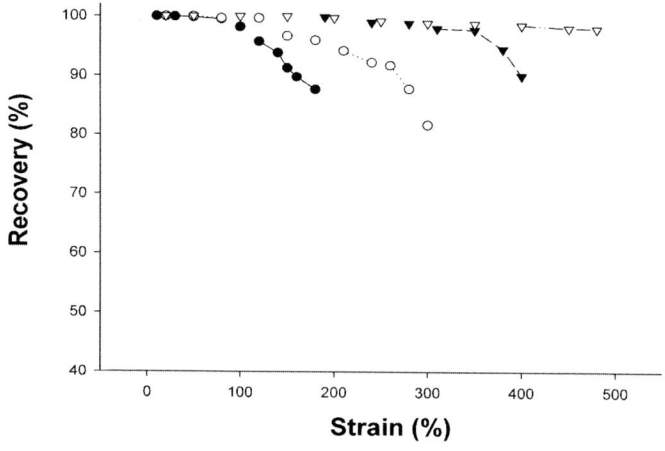

Fig. D.5 Recovery of PLCL scaffolds by gel-pressing method.

D.6 Cautions

1. The scaffolds prepared by the gel-pressing method have proper physical characteristics including a pore size, interconnected pore structure, no skin layer on the outer surfaces, and mechanical strength, which are important properties that affect tissue formation via nutritional diffusion and cell migration.

2. By using the described gel-pressing method, it is possible for the scaffolds to be prepared having optimum mechanical properties and controlled microstructures for several applications in mechanoactive tissue engineering.

E

Protocol of PLGA Microspheres for Tissue-Engineered Scaffold

*Jong Tae Ko,
Jung Su Park,
Byung-Hyun Min and
Gilson Khang*

E.1 Concept

- PLGA-based microspheres are biodegradeable particulate delivery systems providing both drug protection, encapsulated inside a polymeric matrix, and its release at a slow and continuous rate.
- Microsphere manufacturing usually involves (1) the controlling of a disintegrated polymer, (2) cell toxicity, and (3) a suitable environment for cell culture.
- The size and degradeable profile can be easily managed by controlling the molecular weight of the polymer and the process of fabrication.
- PLGA microspheres are particularly attractive for tissue regeneration approaches either as an injectable system or as the integral part of a replacement clinical construct.
- The small, spherical nature of these polymers enables the encapsulation of growth factors or other drugs, and their subsequent delivery to a specific and designated area. Controlled release of bioactive molecules, such as cytokines and growth factors, has become an important aspect of tissue engineering because it allows modulation of cellular function and tissue formation at the afflicted site.
- Cell cultures using microspheres have an advantage of passage abbreviation to improve cell activity.
- The PLGA microspheres manufactured using various methods regulate many aspects of cellular activity, including cell proliferation, cell differentiation, and extracellular matrix metabolism, in a time- and concentration-dependent fashion.

E.2 Procedure

- The polymer is dissolved in a solvent, and is ready to add to a solution in surfactant [Fig. E.1(a)].
- The polymer solution is dropped into an aqueous solution in surfactant by a pipette [Fig. E.1(b)].
- This solution is stirred at 400 rpm for 7 h using a mechanical stirrer [Fig. E.1(c)].

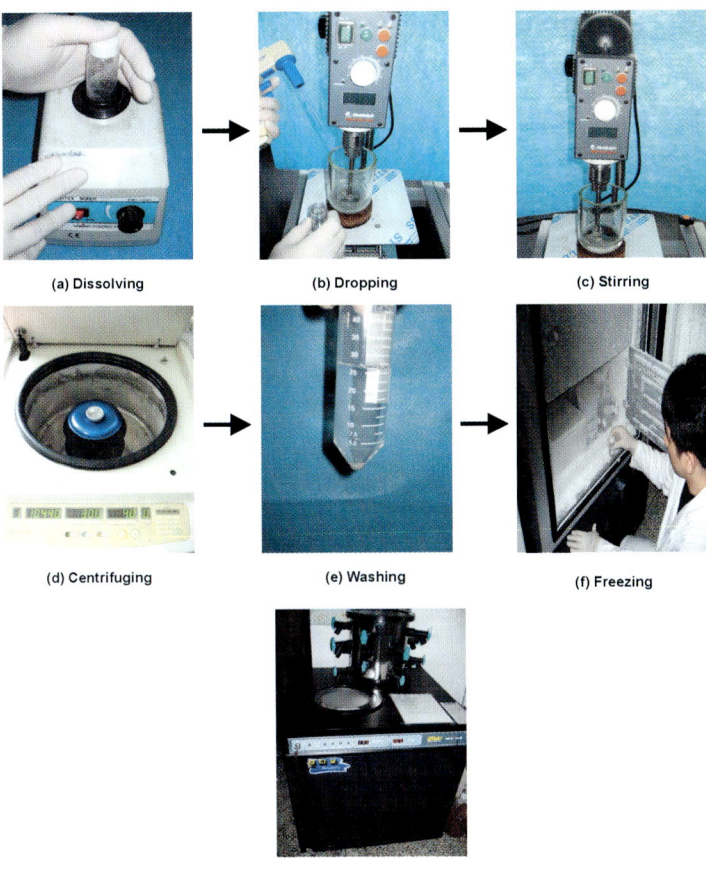

(a) Dissolving

(b) Dropping

(c) Stirring

(d) Centrifuging

(e) Washing

(f) Freezing

(g) Freeze-drying

Fig. E.1 Schematic diagram of the preparation processing of PLGA microsphere scaffolds.

- The fabricated microspheres are collected from the bottom by a centrifugal separator [Fig. E.1(d)].
- The hardened microspheres are centrifuged, washed three times with deionised water, and then kept in a freezer [Figs. E.1(e) and E.1(f)].
- The fabricated microspheres are freeze-dried under low temperature and pressure [Fig. E.1(g)].

E.3 Requirements

1. Polymer
2. Solvent
3. Deionised water
4. Vortex
5. Freeze-dryer
6. Surfactant
7. Stabiliser
8. Centrifugal separator
9. Mechanical stirrer
10. Pipette

E.4 Characterisations

- The sizes of the microspheres are controlled by the concentration of the polymer and the speed of the stirrer.
- The microspheres have a smooth surface and spherical shape (Fig. E.2).
- A cell attached to the microsphere surface produces extracellular matrix (Fig. E.3).
- Cell adhesion and expansion on the microsphere are investigated by SEM. The viablility of attached cells is determined with MTT assay. The expression of special genes with disc cells is evalutated by RT-PCR.

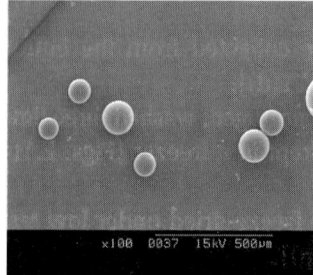 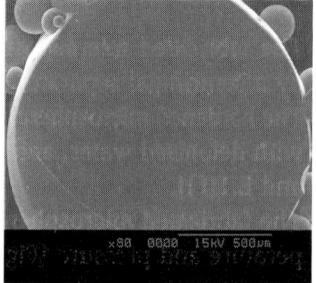

Fig. E.2 Surface and cross-section images of microsphere.

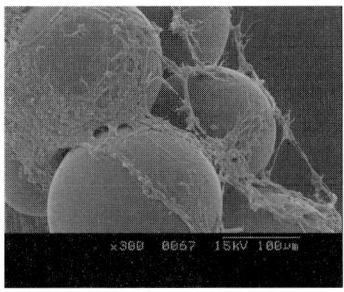 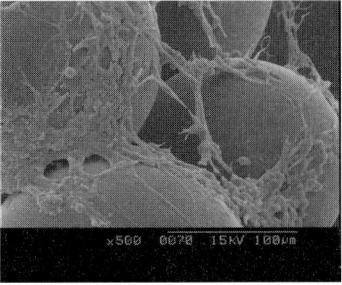

Fig. E.3 Attachment, expansion, and cell aggregation on PLGA microspheres of cultured human disc cells: SEM at 3 days.

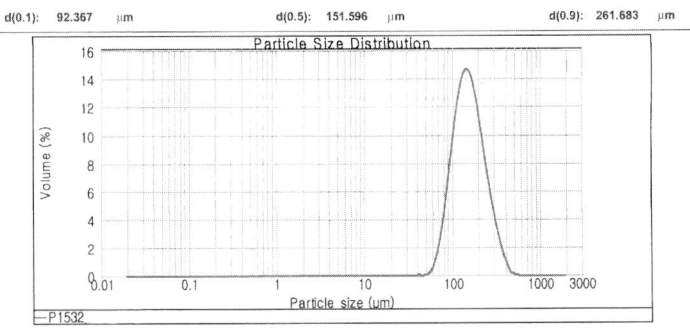

Fig. E.4 The size of microspheres.

- SEM images (Fig. E.3) demonstrate that after 3 days, cells attached to the carriers and the flow intermittency enable aggregation of a number of cell-seeded carriers.
- The distribution of microsphere size ranges from 92 μm to 261 μm, and the average microsphere size is 150 μm (Fig. E.4).

E.5 Cautions

- The concentration of PLGA and the speed of mechanical stirrer play important roles in the fabrication process.

(Continued)

(*Continued*)

- Filtering of the supernatant after the centrifugal separator must be carefully done because of possible loss of the microspheres.
- The stirring time is controlled by the degree of solvent evaporation.

F

Preparation and Usage of Macroporous Bead

Eunhee Bae

F.1 Concept

F.1.1 What is a macroporous bead?

- Macroporous beads are defined as the round-shaped matrices containing uniformly distributed pores larger than 30 μm.
- The materials of macroporous bead for biomedical application should be biocompatible materials such as PLGA, collagen, gelatin, chitosan, hyaluronic acid, cellulose, etc. The macroporous bead has been reported to have advantages in high-density cell culture, due to the extended surface area within a limited space and the improved material exchange rate.
- Therefore, the macroporous bead has drawn attention in tissue engineering fields that require extremely high-density cell culture.

F.1.2 Kinds of macroporous beads

The development of macroporous beads for biomedical application has been carried out in the commercial sector, and plenty of macroporous beads are available in the market (Table F.1).

1. Chitosan is a deacetylated form of chitin that is quantitatively found in the shells of crustaceans (e.g. crabs and shrimps) and insects, and in the cell walls of fungi, mushrooms, and bacteria.
2. It is a polymer consisting of N-acetyl-D-glucosamine repeating units, which are linked to each other via a (1→4)-β-glycosidic linkage. Chitosan is known to be superior in its ability for attaching cells, its biocompatibility, its biodegradeability, and its plasticity to synthetic polymers including PLGA, PLLA, PEG, and PGA.

F.2 Required Equipment

1. Laboratory glassware
2. Magnetic stirrer

(Continued)

Table F.1 Commercially available macroporous beads.

Name	Material	Diameter (μm)	Pore size (μm)	Porosity (%)
Cultispher-G, S, GL (Percell Biolytica)	Gelatin	170–500	~50	50
Informatrix (Biomat Corp.)	Collagen-glycose-aminoglycan	500	40	99
Microsphere (Cellex)	Collagen	500–600	20–40	75
Siran (Schott Glaswerke)	Glass	300–5000	10–400	60
Microporous MC (Solo Hill Labs, Inc.)	Polystyrol	250–3000	20–150	90
Cytopore 1, 2 (Amersham Pharmacia Biotech)	Cellulose	180–210	30	95
Chitopore (REGEN Biotech, Korea)	Chitosan	300–800	30–80	>90
InnoPol (REGEN Biotech, Korea)	PLGA	200–1000	200	>90
Cytoline 1, 2 (Amersham Pharmacia Biotech)	Polyethylene	2000–2500	10–400	65
ImmobaSil ASL	Silicon rubber	1000	50–150	>40

(*Continued*)

3. Syringe pump
4. Teflon-coated plate
5. Freeze-dryer

F.3 Required Reagents

1. Medical grade chitosan MW 70 000–150 000 g/mol
2. Acetic acid
3. Organic solvent (e.g. chloropentane, chloroform, hexane, ethanol
4. NaOH
5. Dry ice

F.4 Procedure

- Dissolve chitosan in 1% aqueous acetic acid solution at a concentration of 1%–2% w/v.
- Dissolve chitosan completely by constantly stirring it over 6 h.
- Filter chitosan solution using 0.45-µm filter paper to remove any undissolved chitosan.
- Prepare EtOH/dry ice bath, and set Teflon-coated plate with organic solvent in the bath.

Note: The temperature of EtOH/dry ice bath can be controlled by adjusting the amount of dry ice.

- Drop chitosan solution into cold (under –5°C) solvent using a syringe pump (Fig. F.1).

Note:

- If a large quantity of macroporous beads is required, any commercially available bead dropper could be used.
- The pore size of macroporous beads is determined by the slope of temperature dropping while the solution is freezing.

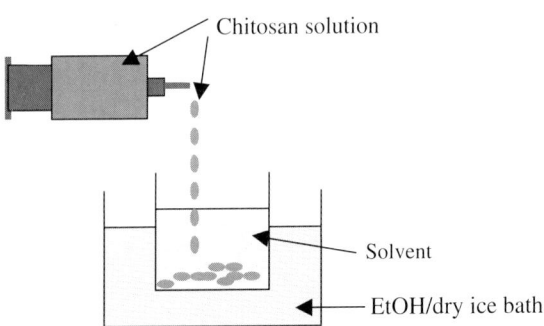

Fig. F.1 Schematic diagram of dropping chitosan solution into the solvent chilled by EtOH/dry ice bath.

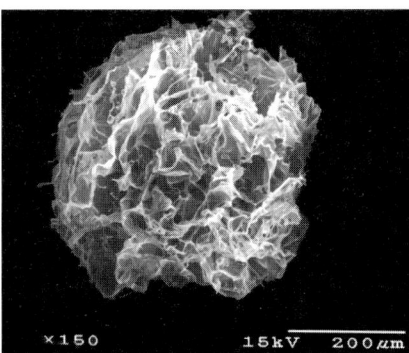

Fig. F.2 SEM of macroporous chitosan bead after neutralisation and second freeze-drying.

- Let the dropped beads completely freeze.
- Transfer the beads to a freeze-dryer after removing the extra solvent.

Note: Take extra care not to thaw the surface of the beads during transfer.

- Freeze-dry the beads for over 24 h.
- Neutralise the dried beads in 1N NaOH in 70% EtOH.
- Wash the neutralised beads five times with distilled water to remove extra NaOH and EtOH.
- Freeze the beads at −70°C and then freeze-dry the beads (Fig. F.2).

F.5 Preparation of Macroporous Beads to be Used in Cell Culture

- Hydrate and swell the dry macroporous chitosan beads in PBS (pH 7–7.5).
- Autoclave the macroporous chitosan beads at 121°C for 15 min.

- After autoclaving, allow the sterilised macroporous chitosan beads to settle, and remove the supernatant.
- Add fresh sterilised PBS and wash the beads twice with sterilised PBS.
- Remove the supernatant, and then add a new culture medium containing serum.
- Leave it overnight at $4°C$.

F.6 Cell Adhesion and Culture into Macroporous Beads

- Seed 500 000–1 000 000 cells per 100-mm culture plate.
- Incubate the cells in a 5% CO_2, $37°C$ incubator until 70%–80% of the plate is covered with cells.
- Wash the cell plate with warm PBS once.
- Add 1 mL of warm trypsin–EDTA solution and incubate for 1–3 min at $37°C$ until the cells become round-shaped.
- Add 3 mL of warm culture medium, and detach cells from the plate by mild pipetting of the medium.
- Put the cell suspension into a 15-mL tube and centrifuge at 1500 rpm for 2 min.
- Aspirate the supernatant.
- Resuspend cells with appropriate culture medium at the desired cell number per unit volume.
- Mix the cell suspension with macroporous beads that were previously prepared to be ready for use. The concentration of chitosan macroporous beads for a 35-mm bacteriological petri dish is approximately 3–6 beads/mL, and the macroporous beads are usually inoculated with 1×10^7 cells/ 3 mL.
- Rock the macroporous beads with cell suspension at $37°C$ under the 5% CO_2 incubator for 4–6 h during the initial culture period.
- Remove the supernatant with unattached cells.
- Add new medium and return the culture plate into the CO_2 incubator with rocking.
- Analyse the cell state using classical techniques such as LDH assay, SEM imaging, etc. (Fig. F.3).

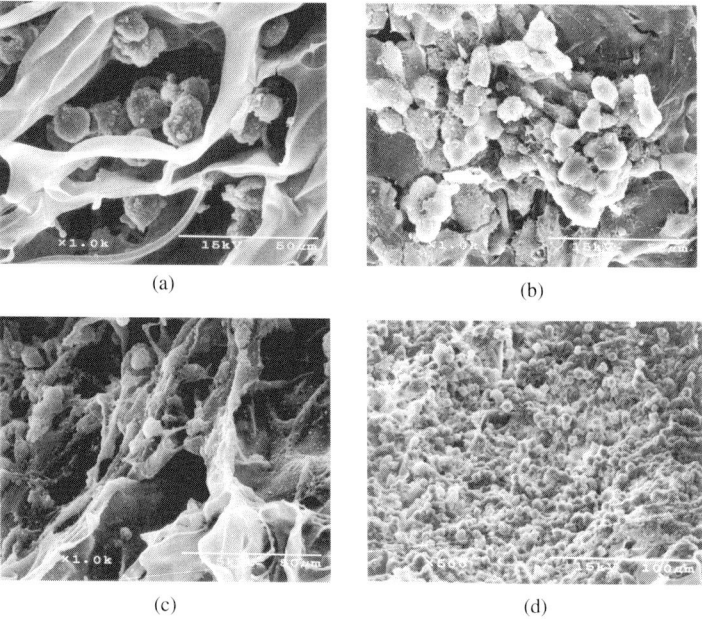

(a)   (b)

(c)   (d)

Fig. F.3 Morphology of various cells attached on macroporous chitosan bead determined by SEM. (a) Rat primary hepatocyte × 1000; (b) CHO × 1000; (c) NIH3T3 × 1000; and (d) PC12 × 500.

> **Note:** Culture conditions using macroporous beads should be adjusted in line with the cell type and culture purpose.

F.7 Examples of Application of Macroporous Beads for Cell Culture and Tissue Engineering

F.7.1 Bioreactor

> 1. Macroporous beads can be used to scale up cultures of adherent cells using various commercially available bioreactors.
>
> *(Continued)*

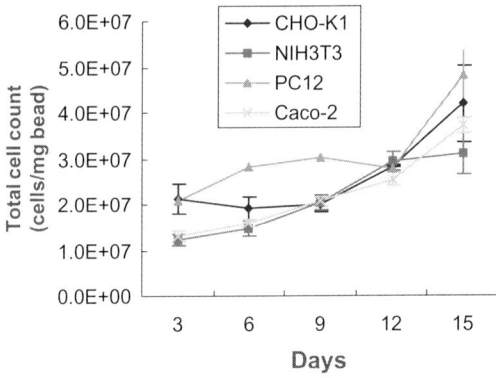

Fig. F.4 The proliferation pattern of various cell lines on macroporous beads determined by NucleoCounter (ChemoMetec, Allerod, Denmark).

(*Continued*)

2. When applying macroporous beads to a bioreactor, mix ready-to-use macroporous beads with cell suspension in as minimum a volume as possible.

3. Control the speed and time of stirring or agitating the cell/bead mixture in order to acquire maximum cell attachment to the surface of macroporous beads either continuously or intermittently.

4. Adjust the volume of culture media and the speed of stirring or agitation for optimum condition (30 rpm is usually suitable for widely used cell lines).

5. Figure F.4 shows the proliferation pattern of various cell lines on chitosan macroporous beads using a spinner flask.

F.7.2 Tissue engineering

1. The applications of macroporous bead as a scaffold for tissue engineering were very limited until recently. Injectable composite tissue-engineering scaffold systems

(*Continued*)

(*Continued*)

are now under development by way of incorporating individual cell carriers within a gel delivery matrix.

2. Differences in physical qualities, namely buoyancy and topography, are dependent on how the pores in macroporous beads are made, and should be customised to provide the optimum conditions for cell–bead interaction and cell functioning in the bead.

3. Any adherent cell type can be cultured and induced to differentiate on the macroporous bead using the culture conditions established for the conventional cell culture system as follows:

   (a) Select a suitable macroporous bead for the desired cell type. Consider the biocompatibility and biodegradeability of the material used for the macroporous bead.

   (b) Attach cells on the surface of the macroporous bead.

   (c) Increase cell numbers using an adequate bioreactor, for example a spinner flask (Bellco, Vineland, NJ, USA), wave reactor (Wave Biotech, Somerset, NJ, USA), or BelloCell (CESCO Bioengineering, Hsin-Chu, Taiwan).

   (d) Induce differentiation by adding factors that could give queue to differentiation.

   (e) Take the cell/bead complex and apply it to *in vivo* animal models either by surgery or injection.

   (f) Determine the efficacy for the reconstitution of the desired tissue using conventional histological methods.

Notes

G

Protocol for Particle-Aggregated Scaffolds

Patrícia B. Malafaya and Rui L. Reis

G.1 Concept

- The scaffolding design (requirements) is a key issue to achieve a successful bone, cartilage, or osteochondral tissue engineering approach.
- A high porosity degree is generally required, but it compromises the scaffolds' mechanical properties. Several scaffold-producing methods include the use of porogen or sacrifice agents that may not assure a high degree of interconnectivity.
- The particle aggregation method described herein allows one to obtain scaffolds with high mechanical properties (thus assuring scaffold stability) and full three-dimensional interconnectivity, which is assured in a 3D perspective by the contact points between the particles.
- The development and tailoring of chitosan-based scaffolds produced by particle aggregation (polymeric, composite, and bilayered) will be presented together with different characterisation techniques in order to achieve a successful model for bone, cartilage, and osteochondral tissue engineering scaffolds.
- The described technique is based on the random packing of prefabricated microspheres with further aggregation by physical or thermal means to create a three-dimensional porous structure.
- The production of polymeric, composite, and bilayered scaffolds for cartilage, bone, and osteochondral applications will be discussed using a simple methodology.
- The production of a biodegradeable template for tissue growth resembling bone and cartilage for use in tissue engineering applications will also be explained.

G.2 Advantages

1. A fully interconnected structure that acts as a scaffold for different tissue engineering applications can be produced.
2. Due to the chitosan bioadhesive properties, the chitosan-based scaffolds produced by particle aggregation present very high mechanical stability.

(Continued)

(Continued)

3. Mechanical stability is assured by particle adhesion and the obtained interface between the particles.
4. Chitosan-based scaffolds with high mechanical properties can be produced.
5. Tailorable scaffolds can be produced. The pore size can be manipulated by controlling the particle size.
6. The methodology is simple.
7. The protocol for scaffold production by means of particle aggregation is described herein with chitosan polymer, but it can be applied with different polymer solutions optimising the particle prefabrication and aggregation principle.

G.3 Procedure

- The chitosan (at desired concentration) is dissolved in acetic acid.
- For the production of composite particles, hydroxyapatite (HA) is added at an adequate concentration to the solution and dispersed homogeneously.
- The chitosan (or chitosan/HA) solution is left overnight to assure complete dissolution [Fig. G.1(a)].
- The chitosan (or chitosan/HA) solution is filtered to eliminate any residual particles.
- The chitosan (or chitosan/HA) solution is extruded through a syringe in a dispenser (syringe pump) at a controlled and constant rate in order to shape the particles into an NaOH solution [Fig. G.1(b)].
- The particle size can be controlled by tailoring the polymer solution concentration, needle diameter, and dispensing rate.
- The particles are then exhaustively washed to remove all exceeding reagents, namely from the precipitation bath [Fig. G.1(c)].
- To produce composite particles, cross-linking can be used with appropriate chitosan cross-linkers. The particles are immersed in the cross-linking solution for a determined short period and then washed again. The cross-linking degree is calculated and optimised in function of the free amine groups.

Fig. G.1 Schematic procedure for the preparation of chitosan-based particle-aggregated scaffolds. (a) Preparation of chitosan solution; (b) set-up of the dispensing syringe system; (c) washing of the produced chitosan particles; (d) production of the chitosan scaffolds in the adequate moulds; and (e) drying of the chitosan particles for further aggregation.

- For the production of chitosan-based scaffolds, the appropriate particles are press-fitted into a specific mould [Fig. G.1(d)] and left in the oven for the necessary time for aggregation to take place [Fig. G.1(e)].
- The chitosan-based scaffolds that have been aggregated are kept until further use (Fig. G.2).

G.4 Requirements

1. Chitosan solution
2. Dispenser (syringe pump)

(*Continued*)

(a)   (b)   (c)

Fig. G.2 Photographs of chitosan-based scaffolds produced by means of particle aggregation. (a) Polymeric scaffolds; (b) composite scaffolds; and (c) bilayered scaffolds.

(*Continued*)

3. Precipitation/complexation bath
4. Mould
5. Oven

G.5 Characterisations

- Chitosan particles produced by precipitation are characterised by a smooth surface and uniform spherical shape with a mean diameter of 500–800 μm (Fig. G.3).
- Chitosan-based scaffolds can be obtained with a mean pore diameter ranging from 100 to 400 μm, with a typical pore morphology as shown in Fig. G.4.
- However, the overall random packing of the chitosan particles into 3D scaffold structures clearly influences the nature of the pores.
- Furthermore, scaffold pore morphology and size can be tailored by controlling the particle size.
- Cross-sections from the bulk of the scaffolds can be stained with eosin and characterised using a stereolight microscope (Fig. G.5).
- In this way, it is possible to further characterise the interface of the particles, showing the chitosan particle bonding that leads

Fig. G.3 SEM microphotograph of prefabricated chitosan-based particles showing their uniformity.

(a)   (b)

Fig. G.4 SEM microphotographs of chitosan-based scaffolds produced by means of particle aggregation. (a) General morphology; and (b) bonding between particles.

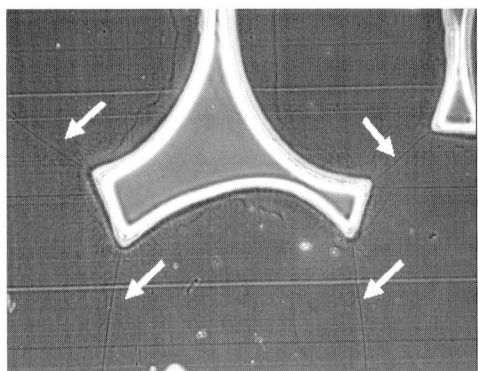

Fig. G.5 Stained cross-section from the bulk of chitosan-based scaffolds observed by stereolight microscopy.

to a very stable interface between the particles and therefore assures the mechanical integrity of the developed scaffolds.

- μ-CT is used to additionally characterise the scaffold morphology and to calculate the porosity of the developed scaffolds through morphometric analysis.
- μ-CT provides an efficient and nondestructive tool to quantitatively measure and qualitatively characterise morphological and morphometric parameters throughout 3D scaffolds.
- To access this characterisation, 2D morphometric analysis can be performed and 3D virtual models can be created using specific software.
- μ-CT is able to accurately evaluate the material (polymeric or ceramic) volume based on 3D spatial distribution.
- X-ray scans should be performed in high-resolution mode. Scan parameters such as power, current, or rotation step should be optimised and maintained constant for all of the scans.
- An adequate number of slices of the materials is chosen and maintained constant in order to analyse the volume of interest of the scaffolds.
- Several types of software can be used as image processing software for CT reconstruction to create and visualise the 3D representation. Examples are NRecon, CT Analyser, CT Vol Realistic 3D Visualisation from SkyScan (Belgium), and Mimics from Materialise (Belgium).
- Two-dimensional morphometric analysis of the scaffolds is performed using an adequate and constant threshold to determine the porosity distribution along the scaffolds (Fig. G.6).
- Three-dimensional virtual models are created using specific image processing software (Fig. G.7).
- One of the main claimed advantages of the described method is the obtained scaffold interconnectivity.
- To assure this property, one may utilise μ-CT analysis.
- The negative of the scaffolds is obtained by using the inverse threshold in order to access the porosity morphology.
- The porosity morphology (Fig. G.8) clearly demonstrates that the described methodology generates scaffolds with a very high degree of interconnectivity.
- In the particular case of bilayered scaffolds for osteochondral tissue engineering, μ-CT is again a very useful morphological and morphometric tool.

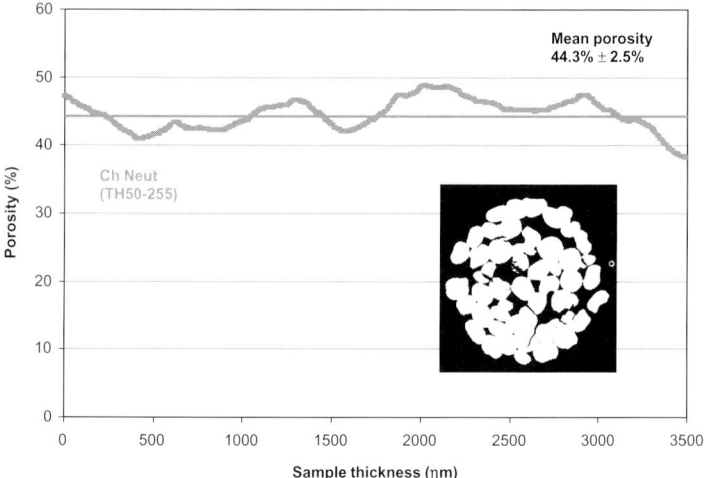

Fig. G.6 Porosity distribution along the polymeric chitosan-based scaffolds, including a binarised image of a single slice of the bulk of the scaffolds.

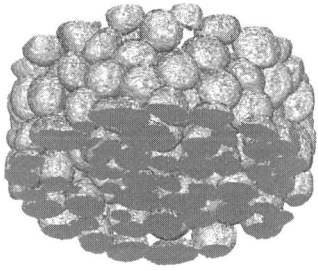

Fig. G.7 3D virtual model of polymeric chitosan-based scaffolds produced by means of the particle-aggregation method.

- With μ-CT, one can characterise the interface between both bony and cartilage components.
- Morphometric analysis is carried out in order to quantitatively characterise the porosity, the porosity distribution, as well as the polymeric and ceramic quantification and distribution along the bilayered scaffolds.
- This analysis clearly allows additional characterisation of the osteochondral scaffolds, as demonstrated in Fig. G.9.

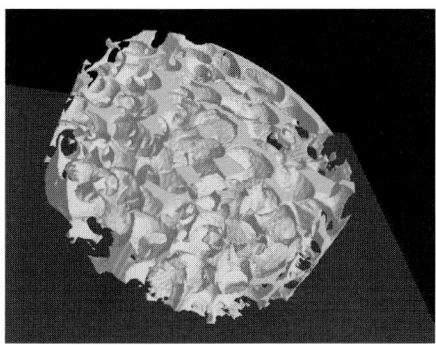

Fig. G.8 3D virtual model of the inverse of polymeric chitosan-based scaffolds showing the porosity morphology.

Fig. G.9 3D virtual model of bilayered chitosan or chitosan/HA-based scaffolds produced by the particle aggregation method.

- Mechanical behaviour and stability are assured by the high adhesion between the particles creating a stable interface.
- To access the mechanical behaviour, compression tests are carried out.
- The mechanical properties of the developed scaffolds are tested on a compressive solicitation mode in a mechanical testing machine in a controlled environment with an adequate cross-head speed.
- The scaffolds show very good mechanical behaviour compared to the typical mechanical properties obtained for chitosan-based porous materials (Fig. G.10).
- For the polymeric scaffolds, a high compressive modulus of 132 ± 7 MPa is obtained.

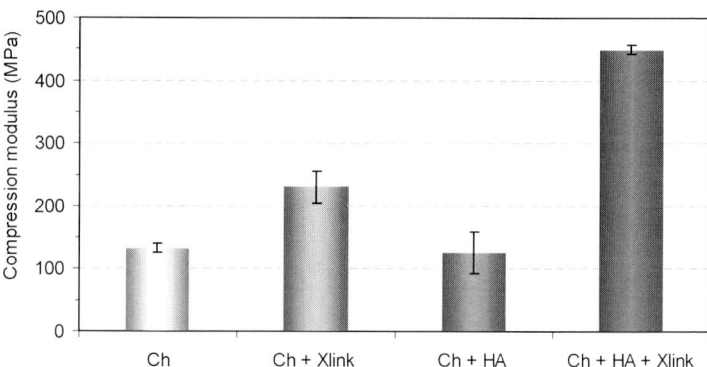

Fig. G.10 Compression modulus of the different produced chitosan-based scaffolds.

- Furthermore, it is important to keep in mind that the mechanical properties may be tailored with scaffold cross-linking or with the incorporation of ceramic fillers (Fig. G.10).

H

Protocol for Freeze-Drying Method

*Chun Ho Kim,
Seung Jae Lee,
Young Ju Choi,
So Hee Yun and
Youngsook Son*

H.1 Concept

- Scaffolds for tissue-engineered medical products have to meet several requirements for successful tissue regeneration: porosity, biodegradeability, and structural integrity.
- In addition, the scaffold has to be durable and flexible in suturing procedure. Porosity can be acquired in the first step of freeze-drying (lyophilising) the solution containing polymeric biomaterials, when the ice crystals of the solvents are formed within the solution.
- The ice crystals serve as a porogen whose size can be easily controlled by adjusting the freezing temperature and the concentration of the solution.
- Structural integrity of the porous sponge-like structure should also be maintained, even after wetting in the interstitial fluid or culture medium.
- If the scaffold is made from water-soluble materials without cross-linkage or if only the 3D frame is maintained by the ionic interaction, then the 3D sponge structure can be easily resolved or turned into a gel-like structure in the aqueous environment.
- This structural integrity in the aqueous environment is determined mainly by the degree of water accessibility within the frames of the scaffold and the presence of degradation enzymes, which is more importantly regulated by the ionic status, water solubility, and innate property of the biomaterial itself.
- Therefore, the concentration of solution, ionic status of biomaterials, solvents, and freezing temperature are important factors to be considered in the manufacturing of suitable tissue engineering scaffolds.

H.2 Chitosan Scaffold

1. Preparation of chitosan solution [Fig. H.1(a)]

   - Dissolve chitosan (Fluka, medium MW; ~700 000) in 1% acetic acid solution (v/v) to give 1.5% solution (w/v) while stirring for 1 h at room temperature.

| (a) Filtrating | (b) Moulding |
| (c) Freezing | (d) Freeze-drying |

Fig. H.1 Preparation processing of scaffold using the freeze-drying method.

- Remove insoluble materials by filtration through sintered glass filter.
- Store the solution overnight at room temperature to remove entrapped air bubbles.

2. Neutralisation of the chitosan solution

- Prepare 1.5% (w/v) chitosan solution in 1% acetic acid, and mix with reconstitution buffer [2.2 g of $NaHCO_3$, 4.77 g of HEPES (200 mM)/100 mL of 0.05N NaOH] and $10 \times$ DMEM/F12 medium without $NaHCO_3$ (DMEM:F12 = 3:1; Gibco BRL) at a ratio of 8:1:1 to obtain a neutralised chitosan solution.

3. Freezing of the solution [Figs. H.1(b) and H.1(c)]

- Pour the chitosan solution into a Teflon-coated mould.
- Freeze the solution at −70°C or −196°C for 12 h.

4. Lyophilisation [Fig. H.1(d)]

- Freeze-dry the frozen solution below 7 mTorr for 48 h.
- Make sure the temperature of the freeze-dryer chamber is below −40°C before transferring the frozen solution to the freeze-dryer.
- Minimise the exposure time while transferring the frozen solution to the freeze-dryer in order to prevent the surface from melting, which may cause closed pores on the surface of the scaffold.

5. Washing (Fig. H.2)

- Remove excess acid within the scaffold by washing with absolute ethanol for 1 h.
- Wash the scaffold serially with 90%, 80%, 70%, 60%, and 50% ethanol for 1 h per each wash.
- Wash the scaffold with water for 3 h.
- Record the residual acid in the final wash by HPLC.

6. Relyophilisation (Fig. H.3)

- Soak the scaffold in the solution containing growth factors or matrix proteins.
- Freeze the scaffold at −70°C and lyophilise the scaffold in order to maintain the original form of the scaffold.

7. Sterilisation

- Sterilise the scaffold either by γ-ray irradiation (5 kGy/5 h; γ-ray source: ^{60}Co) or by EOG.

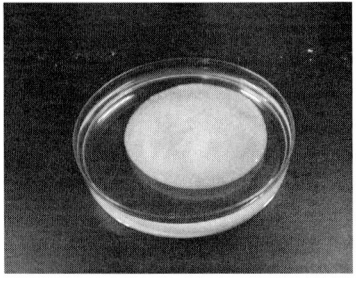

Fig. H.2 Maintenance of the scaffold framework during the washing step.

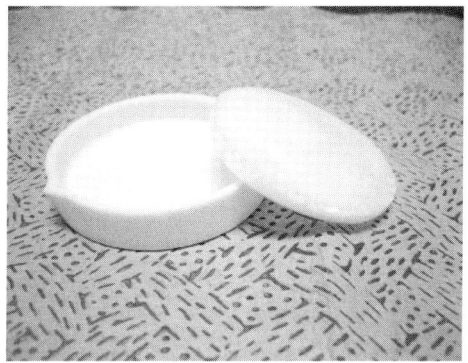

Fig. H.3 The scaffold after relyophilisation.

H.3 Collagen Scaffold or Collagen/ Chitosan-Mixed Scaffold

1. Neutralisation of collagen or collagen/chitosan-mixed solution

   • Neutralise type I p collagen (3 mg/mL pH 3.0; atelomeric porcine collagen) by mixing with reconstitution buffer [2.2 g of $NaHCO_3$, 4.77 g of HEPES (200 mM)/100 mL of 0.05N NaOH] and $10 \times$ DMEM/F12 medium without $NaHCO_3$ at a ratio of 8:1:1, and then mix with neutralised chitosan solution at a proper ratio.

   • All of the procedures should be performed at 4°C to prevent possible gel formation during the procedure.

2. The rest of the procedure can be followed as described above.

H.4 Characterisations (Figs. H.4 and H.5)

• The scaffolds are fractured by a surgical blade and fixed with 1% glutaraldehyde for 1 h at room temperature.
• The scaffolds are dehydrated by immersing them in a series of aqueous solutions of increasing alcohol concentration every 15 min and then freeze-drying them under the same condition as described earlier.

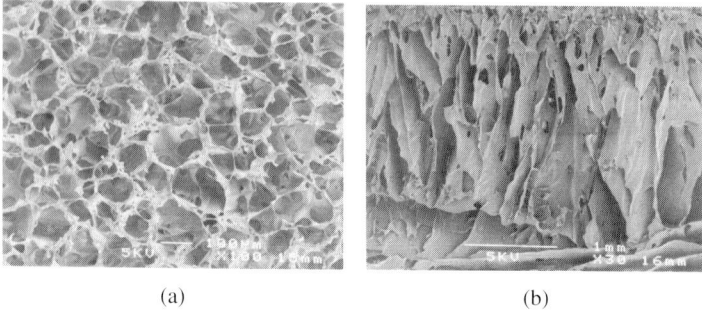

(a)   (b)

Fig. H.4 SEM morphology of the chitosan scaffold. (a) Surface; and (b) cross-section.

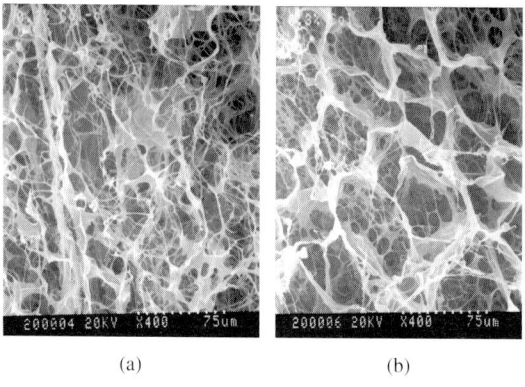

(a)   (b)

Fig. H.5 SEM morphology of the collagen scaffold. (a) Surface; and (b) cross-section.

- The samples are critical-point dried and coated with an ultra-thin gold layer (100 Å). An SEM image is captured by a scanning electron microscope.

Notes

Notes

1

Protocol for Thermally Induced Phase Separation (TIPS)

Hyun Do Kim and Doo Sung Lee

I.1 Concept

- Freeze-drying via TIPS has received much attention in industrial applications for the production of isotropic, highly interconnected, and porosity-designed membranes.
- Figure I.1 represents a schematic temperature–composition phase diagram for a binary polymer/solvent system.
- Above the binodal curve, a single polymer solution phase is formed; and if cooling below the curve, polymer-rich and polymer-poor phases are separated in a thermodynamic equilibrium state.
- The spinodal curve is defined as the line at which the second derivative Gibbs free energy of mixing is equal to zero, and it divides the two-phase region into unstable and metastable regions.
- If the system is quenched into the metastable region, phase separation occurs in a nucleation and growth mechanism, leading to a bead-like isolated cellular structure.

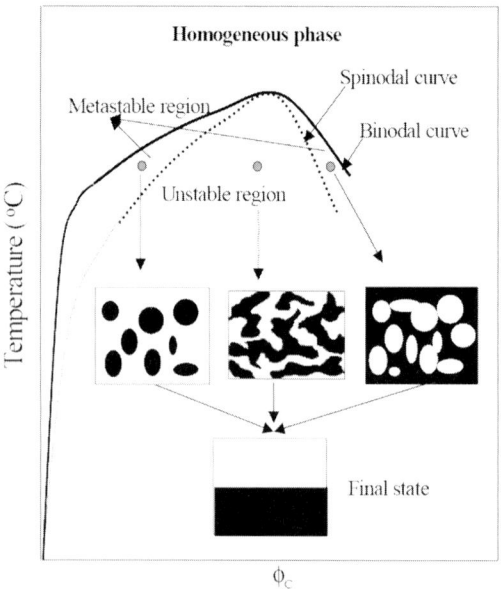

Fig. I.1 Schematic representation of a typical polymer-solvent-nonsolvent ternary phase diagram.

- On the other hand, if the system temperature is quenched into the unstable region, phase separation takes place in a spinodal decomposition mechanism (liquid–liquid phase separation), resulting in a microporous interconnected structure.
- The phase separation and freeze-drying method appears as a versatile technique to prepare highly porous three-dimensional polymer scaffolds that fulfil all of the requirements for cell transplantation.
- Porosity can be controlled in terms of pore size and morphology by a suitable choice of processing conditions and by a strict control of phase separation conditions such as quenching temperature, quenching depth, ageing time, polymer concentration, molecular weight, solvent/nonsolvent composition, and additives.

I.2 Phase Diagram

- The cloud-point curve of the PLLA ternary system is determined by visual turbidimetry (Table I.1).
- Weighted PLLA (1, 3, 4.5, 5.5, or 7 wt%) — with or without PEG/PEG–PLLA diblock copolymers — is added into a 4-mL vial tube with 1,4-dioxane/water mixture (87/13, w/w) as a solvent.
- This is dissolved by heating at 65°C for 5 h with a magnetic stirrer.
- The homogeneous PLLA solution is reheated to ca. 10°C above the expected cloud-point temperature (55°C).
- In water bath, the solution is slowly cooled in steps of 1°C, equilibrating the system for 10 min at each new temperature.
- The cloud point is reported as the temperature at which the clear solution becomes turbid by visual sight (Fig. I.2).

Table I.1 PEG, PEG–PLLA diblock copolymers.

	Copolymer	$Mn/g \ mol^{-1}$
MPEG5000	PEG_{114}	5000
Diblock1	PEG_{114}–$PLLA_6$	5000–413
Diblock2	PEG_{114}–$PLLA_{40}$	5000–2845
Diblock3	PEG_{114}–$PLLA_{62}$	5000–432
Diblock4	PEG_{45}–$PLLA_{26}$	2000–1830

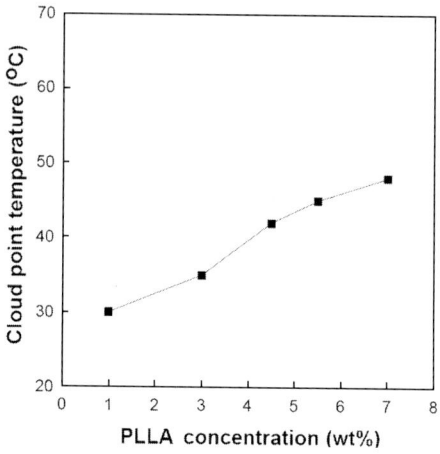

Fig. I.2 Cloud-point curve on PLLA-dioxane-water system (PLLA, Lacty 5000; Mw, 218 000 g/mol; PDI, 1.55).

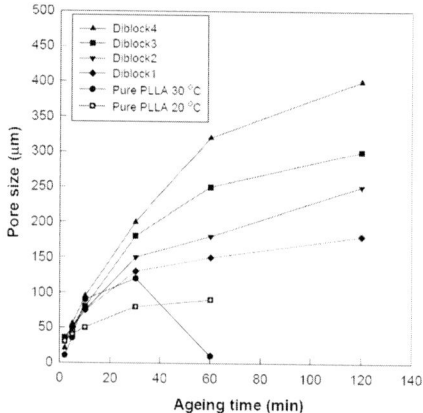

Fig. I.3 Pore size distribution of PLLA scaffolds as a function of ageing time. ● Pure PLLA; ◆ added diblock1 0.5 wt%; ▼ added diblock2 0.5 wt%; ■ added diblock3 0.5 wt%; ▲ added diblock4 0.5 wt% at a quenching temperature of 30°C; and □ pure PLLA at a quenching temperature of 20°C.

I.3 Preparation of PLLA Scaffold

The pore size of scaffolds must be optimised according to cells or tissue. In the PLLA-dioxane-water ternary system, the pore size of PLLA scaffolds is increased as a function of the ageing time at a quenching temperature of 30°C (Fig. I.3).

I.4 TIPS and Freeze-Drying Method (Fig. I.4)

- PLLA (4.5 or 5.5 wt%) solution with a mixture of 1,4-dioxane and water (87/13 w/w) as the solvent is prepared.
- The sample is reheated to 15°C above the measured cloud-point temperature, and then placed in a water bath preheated to the quenching temperature.
- It is kept for 2, 10, 30, 60, or 120 min at the quenching temperature.
- The annealed sample is directly immersed in liquid nitrogen to be fast-frozen for 1 h, and then a small hole is cut in the vial cap to release the solvent.
- Freeze-drying is performed in a freeze-dryer at −77°C and 7 mTorr for 3 days in order to remove the solvent and obtain the macroporous scaffolds (Fig. I.5).

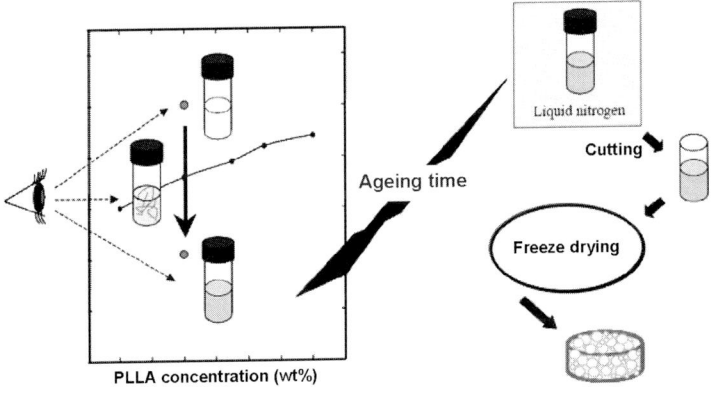

Fig. I.4 Fabrication of PLLA scaffold via TIPS method.

Fig. I.5 PLLA solution quenching.

- The dry scaffolds are cut into cubes with a surgical blade (7–8 mm, thickness 2 mm; 13–15 mg PLLA).
- Prior to cell seeding, 3D scaffolds are prewetted with 70% ethanol for 3 h to sterilise them and enhance their water uptake.
- The ethanol is removed by soaking with agitation for 1 h in six changes of PBS, and then the scaffolds are left overnight in the culture media.
- The characterisation of scaffolds is performed by SEM (Fig. I.6).

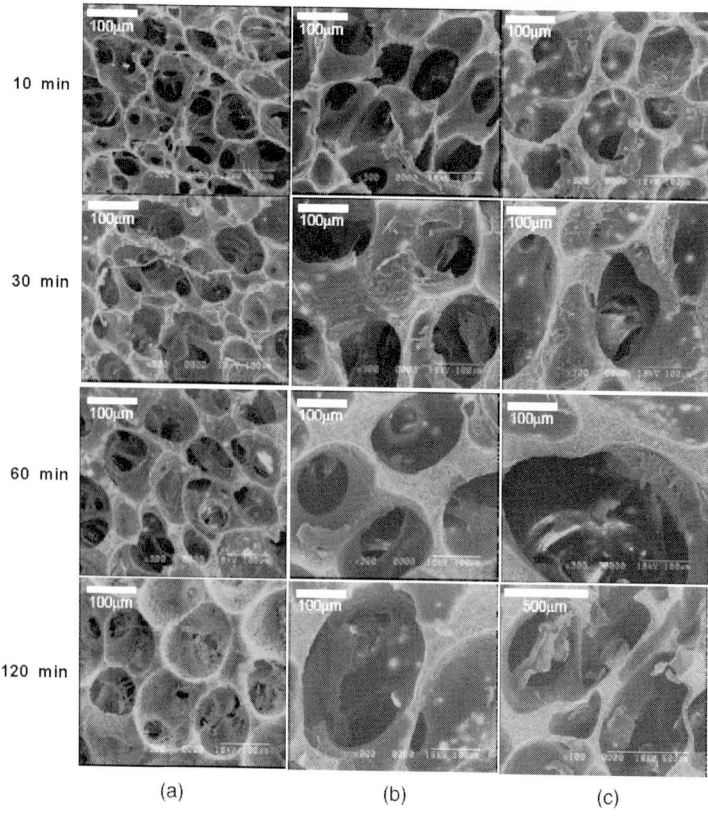

Fig. I.6 Pore size of scaffolds prepared from a 4.5 wt% PLLA-dioxane-water (87/13, w/w) solution as a function of ageing time at a quenching temperature of (a) 25°C, (b) 30°C, and (c) 35°C.

I.5 Requirements

1. PLLA
2. 1,4-dioxane/water (87/13, w/w)
3. PEG, PEG–PLLA diblock copolymers
4. Liquid nitrogen
5. 70% ethanol
6. PBS
7. Culture media
8. 4-ml vial
9. Scales
10. Magnetic stirrer
11. Water bath and temperature controller
12. Timer

J

Protocol for Centrifugation Method

Se Heang Oh and Jin Ho Lee

J.1 Concept

- Recently, a centrifugation method has been introduced as an effective method to fabricate scaffolds that have various shapes with a uniform surface and inside pore structures (Fig. J.1).
- The centrifugation method is very effective for preparing scaffolds with complicated shapes for tissue engineering applications.
- The scaffolds can be fabricated in various shapes from many different natural and synthetic polymers by the centrifugation method (Fig. J.2).

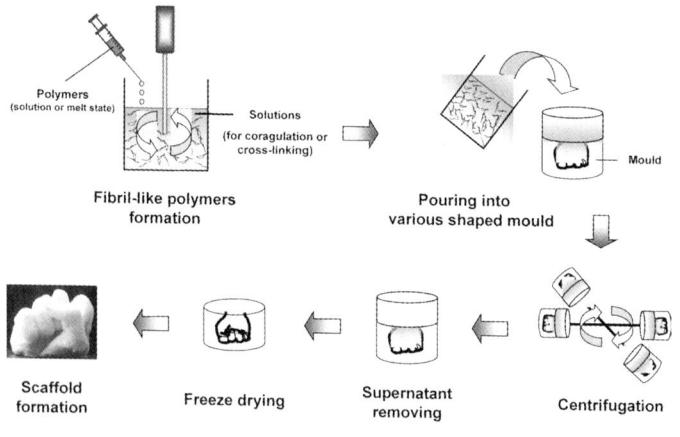

Fig. J.1 Schematic diagram showing the fabrication of scaffolds by a centrifugation method.

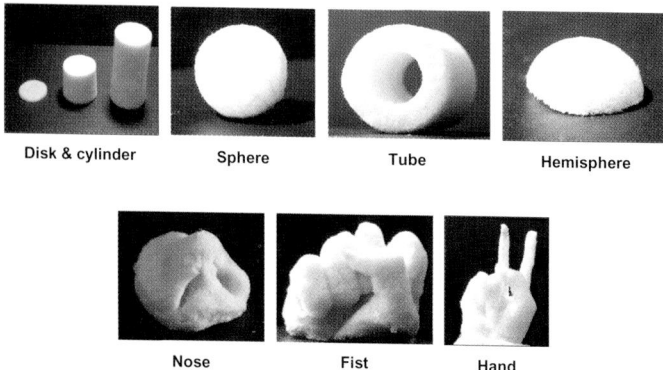

Fig. J.2 Photographs of variously shaped scaffolds fabricated by the centrifugation method.

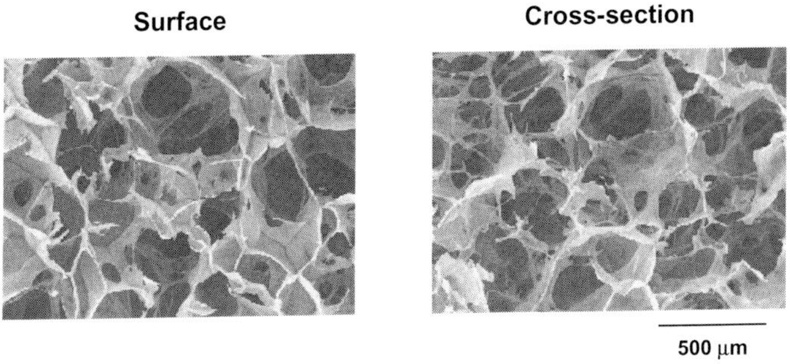

Surface   **Cross-section**

500 μm

Fig. J.3 SEM photographs of a chitosan scaffold fabricated by the centrifugation method (centrifugation speed, 3000 rpm; ×100).

J.2 Chitosan Scaffold

- Prepare 2 wt% chitosan solution (in 0.1 M acetic acid) and pour into a 10-mL syringe.
- Prepare 0.05 M NaOH aqueous solution in a 500-mL beaker.
- Slowly drop the chitosan solution into the NaOH solution with vigorous agitation using a homogeniser.
- Obtain the fibril-like chitosan precipitate (suspended in NaOH solution).
- Wash the fibril-like chitosan in excess phosphate buffered saline solution (PBS, pH ~7.4) and the following distilled water to remove residual acetic acid and NaOH.
- Obtain neutralised fibril-like chitosan suspension [in distilled water (pH ~7.0)].
- Place the fibril-like chitosan-suspended solution in a cylindrical (or various-shaped) mould.
- Centrifuge at 3000 rpm for 5 min for the fibril-like chitosan accumulation in the mould and the following fibril bonding.
- Discard supernatant from the mould.
- Freeze the fibril-like chitosan accumulation in the mould at −70°C for 12 h and then lyophilise it.
- Obtain the cylindrical (or various-shaped) scaffold (Fig. J.3).

J.3 Alginate Scaffold

- Prepare 2 wt% sodium alginate aqueous solution and pour into a 10-mL syringe.
- Prepare 2 wt% CaCl$_2$ aqueous solution in a 500-mL beaker.

- Slowly drop the sodium alginate aqueous solution into the CaCl$_2$ solution with vigorous agitation using a homogeniser.
- Obtain the fibril-like Ca–cross-linked alginate (suspended in CaCl$_2$ solution).
- Wash the fibril-like alginate in excess distilled water to remove residual CaCl$_2$.
- Obtain the fibril-like alginate suspension [in distilled water (pH ~7.0)].
- Place the fibril-like alginate-suspended solution in a cylindrical (or various-shaped) mould.
- Centrifuge at 3000 rpm for 5 min for the fibril-like alginate accumulation in the mould.
- Discard supernatant from the mould.
- Freeze the fibril-like alginate accumulation in the mould at −70°C for 12 h and then lyophilise it.
- Obtain the cylindrical (or various-shaped) prescaffold.
- Immerse the prescaffold in 1 wt% chitosan solution (in 0.1 M acetic acid) for 30 min to prevent disentanglement of fibrils in cell culture medium or physiological solution.
- Wash the prescaffold in excess distilled water to remove residual acetic acid and chitosan.
- Freeze the chitosan-coated alginate prescaffold at −70°C for 12 h and then lyophilise it.
- Obtain the cylindrical (or various-shaped) scaffold (Fig. J.4).

J.4 PCL Scaffold

- Fill a 100-mL beaker with PCL powder and Pluronic F127 powder (1/1.5, w/w) (Pluronic F127 helps to reduce the viscosity

1000 rpm
(~250 μm*)

2000 rpm
(~190 μm*)

3000 rpm
(~130 μm*)

500 μm

Fig. J.4 SEM photographs of an alginate scaffold fabricated by the centrifugation method (*, pore size; ×100).

of melted PCL, and also provides the hydrophilicity in the prepared PCL scaffold).

- Heat it to 120°C and mix it using a homogeniser until a homogeneous molten state is reached.
- Slowly drop the viscous PCL/Pluronic F127 mixture into distilled water (at room temperature) with vigorous agitation using a homogeniser.
- Obtain the fibril-like PCL precipitate (suspended in distilled water).
- Wash the fibril-like PCL in excess distilled water to remove residual Pluronic F127.
- Obtain the fibril-like PCL suspension [in distilled water (pH ~7.0)].
- Place the fibril-like PCL-suspended solution in a cylindrical (or various-shaped) mould.
- Centrifuge at 3000 rpm for 5 min for the fibril-like PCL accumulation in the mould.
- Discard supernatant from the mould.
- Freeze the fibril-like PCL accumulation in the mould at −70°C for 12 h and then lyophilise it.
- Obtain the cylindrical (or various-shaped) prescaffold.
- Heat the prescaffold up to almost PCL melting temperature for fibril bonding in the prescaffold.
- Obtain the cylindrical (or various-shaped) scaffold.

J.5 Requirements

1. PCL
2. Pluronic F127
3. Chitosan
4. Acetic acid
5. NaOH
6. Sodium alginate
7. $CaCl_2$
8. High-speed homogeniser
9. Magnetic stirrer/heater

(Continued)

(*Continued*)

10. Beakers
11. Magnetic stirring bar
12. Spatula
13. Cylindrical (or variously shaped) mould
14. Centrifuge
15. Deep freezer
16. Freeze dryer
17. PBS
18. Distilled water

J.6 Characterisations

The pore size of the scaffold can be controlled by adjusting the centrifugal force (i.e. increasing centrifugal force leads to the fabrication of scaffolds with smaller pore size) (Fig. J.4).

K

Injectable Thermosensitive Gel

Soo Chang Song

K.1 Concept

K.1.1 Injectable polymeric systems

- One of the simplest and most convenient approaches in tissue engineering applications is to inject the polymer–cell or polymer–drug entity into the body.
- Injectable systems offer specific advantages over preformed scaffolds, including easy application, site-specific delivery, and improved compliance and comfort for patients.
- Water-soluble, thermosensitive, or pH-sensitive polymers exhibiting reversible sol–gel transition and photopolymerisable hydrogels have been tailor-made as injectables.

K.1.2 Thermosensitive hydrogels

- Thermosensitive hydrogels can be formed either by physical gelation without covalent bonding (e.g. ionic interaction, hydrophobic association, hydrogen bonding between polymer chains in an aqueous solution) or by chemical gelation caused by thermosensitive chemical cross-linkers.
- The former may go through sol–gel phase transitions in response to changes in temperature, but the latter may undergo swelling/shrinking.
- Thermosensitive hydrogels made by physical cross-links between polymer chains are very useful for injectable tissue engineering because no toxic organic cross-linkers are usually employed.

K.1.3 Polyphosphazene

- Polyphosphazenes are a new class of inorganic backbone polymers that are superior to many other organic systems in terms of their molecular structural diversity and property variations.
- These polymers can be used as a reactive macromolecular intermediary by replacing chlorine atoms with organic side groups to give various hydrolytically stable polymers.

K.2 Procedure

$m = 7, 11, 16$

NHR =

IleOEt

NHR' = $NHCH_2COOCHCOOCH_2CH_3$
                              |
                             CH_3

GlyLacOEt

- Before the reaction, L-isoleucine ethyl ester (IleOEt), glycolic or lactic acid ester, and α-amino-ω-methoxy-PEG (AMPEG) are respectively dried for 1 day at 50°C in vacuum for moisture removal.
- THF is dried by reflux over sodium/benzophenone under nitrogen atmosphere.
- TEA and acetonitrile are distilled over BaO under nitrogen atmosphere.
- L-isoleucine ethyl ester hydrochloride suspended in dry THF-containing triethylamine is slowly added to poly(dichlorophosphazene) dissolved in dry THF.
- The reaction is performed for 4 h at 4°C, and then for 20 h at room temperature.
- TEA and ethyl-2(O-glycol)lactate (GlyLacOEt) oxalic salt dissolved in acetonitrile are added to this mixture, and the reaction mixture is stirred for 19 h at room temperature.
- After AMPEG dissolved in dry THF-containing TEA is added to the polymer solution, the reaction mixture is stirred for 2 days at 40°C–50°C.
- The above reaction mixture is filtered.
- After the filtrate is concentrated, it is poured into n-hexane to obtain precipitate, which is reprecipitated twice in the same solvent.

- The reprecipitated polymer is concentrated.
- The polymer product is further purified by dialysis in methanol for 4 days and then in distilled water for 4 days at 4°C.
- The final dialysed solution is freeze-dried to obtain the final polymer.

K.3 Gelation Properties of Polymers (Fig. K.1)

- Most of the poly(organophosphazenes) synthesised in this work show sol–gel transition properties in aqueous solution.
- The gelation properties of the polymers are dependent on several factors, such as the composition and size of the substituents, the amino acid esters used, and the polymer conditions.
- The gelation of the polymers is imparted mainly by hydrophobic associations of hydrophobic groups of amino acid esters.

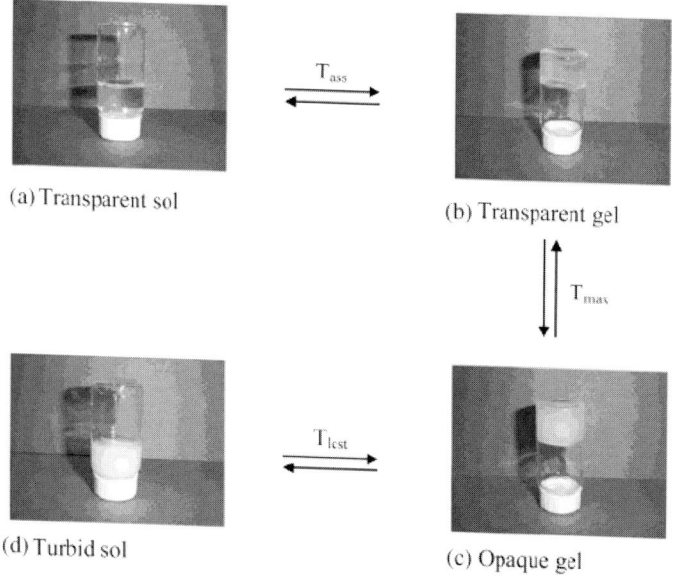

(a) Transparent sol

T_{ass}

(b) Transparent gel

T_{max}

(d) Turbid sol

T_{lcst}

(c) Opaque gel

Fig. K.1 Photographs of the sol–gel transitions of poly(organophosphazenes) in this study in aqueous solution, observed with a gradual temperature increase from (a) to (d).

K.4 Isolation of Rabbit Chondrocyte

- Sacrifice a rabbit after general anaesthesia.
- Shave the hair around the knee.
- Dissect the knee under sterile condition [Fig. K.2(a)].
- Cut out cartilage pieces from the joint surface.
- Digest the small slices of cartilage at 37°C with 0.2% type II collagenase for 1 h in DMEM [Fig. K.2(b)].
- After the completed digestion, filter the cell suspension through a nylon mesh filter (pore size, 60 μm).
- Before seeding, wash the chondrocytes with PBS.
- Seed the chondrocytes onto polystyrene culture dishes (10 cm in diameter), and grow in culture until a confluent monolayer has formed with DMEM (10% FBS, 1% penicillin/streptomycin) [Fig. K.2(c)].
- Perform cell culture under standardised conditions (37°C, 5% CO_2, saturated humidity) in an incubator.
- Change the culture medium every 2–3 days.

K.5 Preparation of Gel Mixture and Subcutaneous Injection into Nude Mouse

- Dissolve poly(organophosphazene) in cold culture medium.
- Stir polymer solution at 4°C until the solution is homogeneous [Fig. K.2(d)].
- Sterilise the melted solution with UV for 2 h.
- Mix the sterilised polymer solution of 200 μL, prepared rabbit primary chondrocytes, and signalling molecule (TGF-β) [Fig. K.2(e)].
- Inject the prepared polymer mixture subcutaneously into the nude mouse [Fig. K.2(f)].
- After several weeks, analyse the injected chondrocytes.

K.6 Requirements

1. Hexachlorocyclotriphosphazene
2. Aluminium chloride

(Continued)

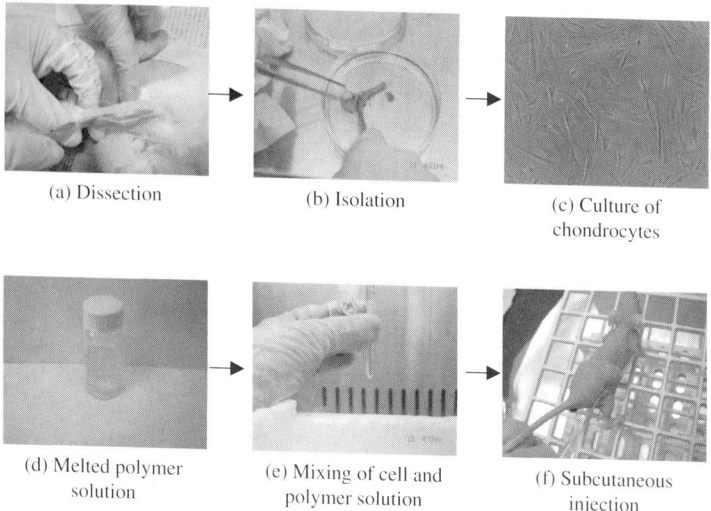

(a) Dissection

(b) Isolation

(c) Culture of chondrocytes

(d) Melted polymer solution

(e) Mixing of cell and polymer solution

(f) Subcutaneous injection

Fig. K.2 Protocol for implantation of chondrocytes/polymer solution mixture.

(*Continued*)

3. Sublimator [Fig. K.3(a)]
4. Glove box [Fig. K.3(b)]
5. High-temperature oven
6. Glassware
7. Prepolymer
8. IleOEt
9. GlyLacOEt
10. AMPEG (Mw = 350, 550, 750)
11. Dried THF
12. Dried TEA
13. Dried acetonitrile
14. Hexane
15. Methyl alcohol
16. Distilled water
17. 250-mL flask
18. 500-mL flask

(*Continued*)

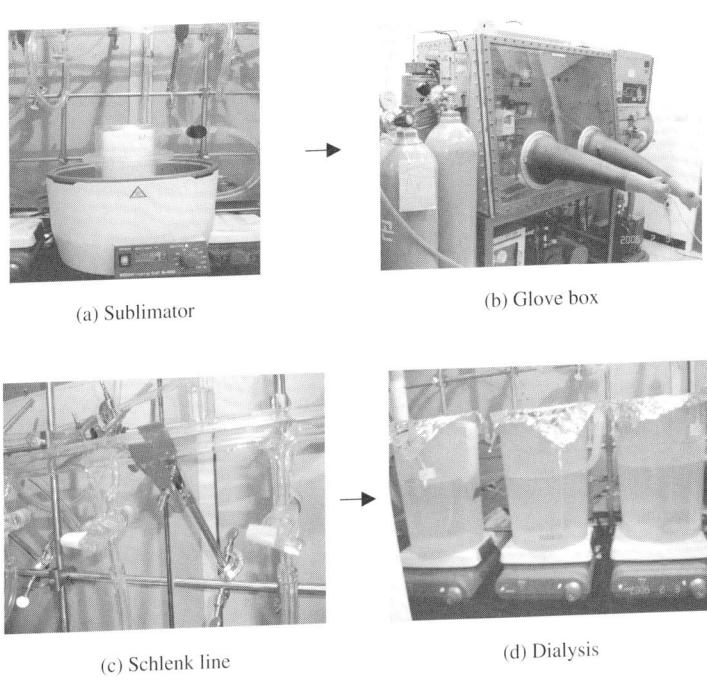

(a) Sublimator

(b) Glove box

(c) Schlenk line

(d) Dialysis

Fig. K.3 Experimental setting for polymerisation process of poly(organophosphazene).

(Continued)

19. 1000-mL flask
20. Schlenk line [Fig. K.3(c)]
21. Magnetic stirrer
22. Cannula
23. Heating mantle
24. Dialysis membrane tubes (Mw = 12 000–14 000) [Fig. K.3(d)]
25. Poly(organophosphazenes)
26. 0.01M PBS aqueous solution pH 7.4
27. Viscometer
28. Syringe

(Continued)

(*Continued*)

29. PBS
30. Type II collagenase
31. DMEM (10% FBS, 1% penicillin/streptomycin)
32. CO_2 incubator
33. Operating scissors and knife
34. Rabbit (2 weeks old)
35. Trypsin–EDTA
36. Trypan blue
37. Hemocytometer
38. TGF-β

L

Preparation of an Acellular Scaffold

Sang-Soo Kim and Byung-Soo Kim

L.1 Concept

To engineer tissues successfully, the selection of scaffolds is critical. Although various synthetic biodegradeable polymer scaffolds have been developed and improved by mimicking biological structures, acellular scaffolds may be a better option for the following reasons:

- Acellular scaffolds retain their correct anatomical structure even after the decellularisation process.
- Acellular scaffolds retain native ECM architecture and possess the cell adhesion ligands.
- The decellularisation process considerably reduces immunological responses by completely removing cellular components.
- The decellularisation process facilitates similar biomechanical properties as those of native tissues that are critical for the long-term functionality of the grafts.
- The decellularisation process facilitates good handling characteristics.

L.2 Procedure

- Carefully excise tissues, and remove adherent fat and other connective tissues.
- Immediately wash the isolated tissues in PBS solution.
- Immerse isolated tissues in distilled water at 4°C for 24 h with continuous shaking at 200 rpm. Replace the distilled water several times during this period (Fig. L.1).
- Treat the tissues with decellularising solution in a 500-mL flask, and incubate at 4°C for 72 h for cell lysis with continuous shaking at 200 rpm. Replace the decellularising solution every 24 h.
- Wash the tissue thoroughly with distilled water several times to remove residual detergents. Repeat the third step.
- Lyophilise the treated tissues for 24 h (Fig. L.2).
- Sterilise with cold EOG and store at room temperature.

Fig. L.1 Tissues immersed in the decellularising solution with continuous shaking at 200 rpm using an orbital shaker.

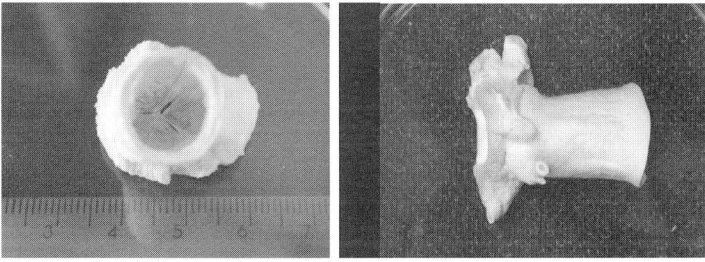

Fig. L.2 Macroscopic images of decellularised porcine pulmonary valve.

L.3 Requirements

1. 500-mL flask
2. Orbital shaker
3. Lyophiliser
4. Ethylene oxide sterilisation unit
5. Triton X-100 (Sigma, St. Louis, MO)
6. Ammonium hydroxide
7. PBS
8. Distilled water
9. Decellularising solution [0.5% (v/v) Triton X-100 and 47.6 mM ammonium hydroxide in distilled water].

L.4 Characterisations

- The microstructure and ECM architecture of decellularised scaffolds can be examined using SEM (Fig. L.3).
- Acellularity and the preservation of native ECM architecture of the decellularised scaffolds can be confirmed with histological analyses (Fig. L.4).
- Acellularity of the decellularised scaffolds can be confirmed with DNA quantification analysis (Fig. L.5).

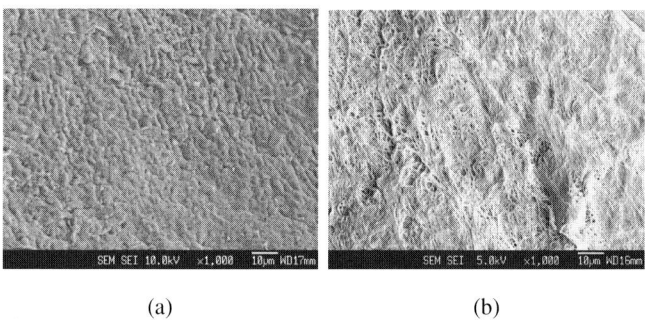

(a)   (b)

Fig. L.3 SEM image of (a) a native and (b) a decellularised porcine pulmonary valve. The endothelial layer was removed during the decellularisation process.

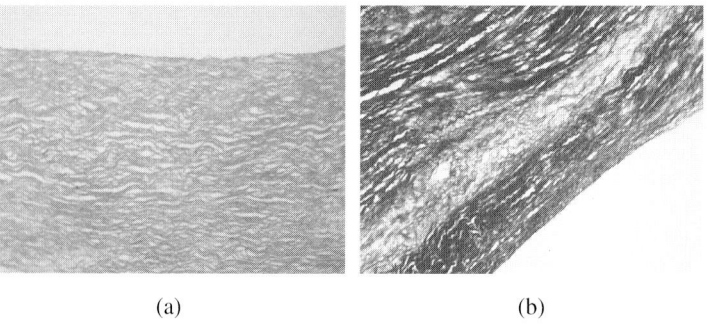

(a)   (b)

Fig. L.4 Histological analyses of a decellularised porcine pulmonary valve. Histological analyses of acellular scaffolds show complete removal of the cells and preservation of the native ECM (collagen) architecture. (a) Hematoxylin and eosin staining; (b) Masson's trichrome staining.

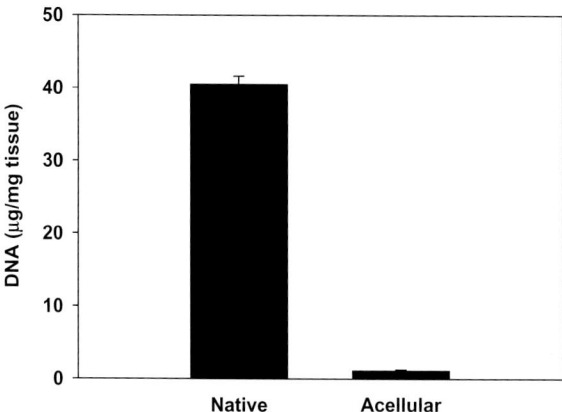

Fig. L.5 DNA assay of acellular scaffolds confirms the nearly complete removal of cellular components from the tissues. DNA quantification shows a reduction of more than 97% in the DNA content of acellular scaffolds as compared with normal tissues.

L.5 Cautions

Leave the sterilised scaffolds for a sufficient period (>48 h) before use for the release of EOG.

M

Protocol for Self-Assembled Human Hair Keratins

Sang Jin Lee and Mark van Dyke

M.1 Concept

- Keratins are an important biocompatible material for human medical applications for two main reasons: (1) they can be extracted as soluble proteins from readily available allogenous tissue, namely end-cut hair fibre, so they are inexpensive to obtain; and (2) keratins contain sites of molecular recognition where integrin-mediated cell attachment can occur.
- Developing useful biomaterials from keratins is facilitated by chemical methods that permit the extraction of soluble proteins from human hair fibres.
- The two main chemistries for solubilising keratins are oxidation and reduction.
- Oxidative cleavage results in the formation of keratoses, and reductive cleavage results in the formation of kerateine analogues that contain cysteine residues.
- Keratins can be of low (so-called "gamma," ca. 15 kDa) or high (so-called "alpha," ca. 85 kDa) molecular weight and contain differing amounts of cysteine, depending on their location in the follicle.
- Keratin biomaterials have been investigated in a plethora of medical applications including drug delivery, wound healing, soft tissue augmentation, synthetic skin, coatings for implants, and scaffolds for tissue engineering.
- The biocompatibility of these constructs has been tested and, in many instances, has been found to exceed that of other naturally derived materials. In fact, cell adhesion and growth on keratin substrates, an important characteristic for tissue engineering, is excellent.
- A recent publication suggests that the cellular recognition of keratins is due in part to a molecular mechanism made possible by the numerous cell adhesion sequences found on keratin proteins.

M.2 General Hair Preparation (Fig. M.1)

- Wash hair in warm water with a mild detergent and rinse thoroughly with deionised (DI) water.
- After drying, mince hair manually using sharp scissors into pieces not exceeding approximately 1 inch in length.

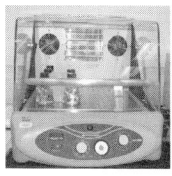
Vigorous shaking

Screening to remove the hair

Filtration

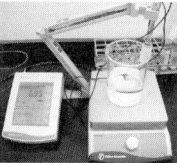
pH adjustment by titration

Centrifugation

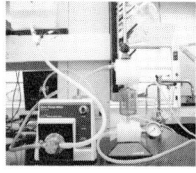
Dialysis

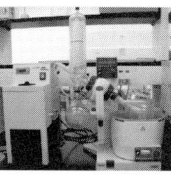
Concentration using rotary evaporator

Lyophilisation

Fig. M.1 Equipment used for keratin biomaterials processing.

- Extract hair in hexane by placing in a closed plastic container and shaking thoroughly. Repeat the hexane wash two more times.
- Repeat the triplicate washes again with absolute ethanol.
- Air-dry the hair in a chemical fume hood overnight.

M.3 Isolation of Keratoses

- Extract the keratoses by a two-step process wherein the cystine bonds are broken down by oxidation in the first step.
- Prepare an 8.0-w/v% solution of peracetic acid in DI water (20-fold excess).
- Perform the oxidation in a plastic container with sufficient volume to allow for vigorous mixing. Seal the container and agitate at 4°C for 24 h (or at 37°C for 12 h).
- Allow the solution to equilibrate to room temperature.
- Remove hair from the liquid by passing the solution through a sieve (425 μm).
- Remove free keratoses from the oxidised hair fibres by repeated extraction with a strongly denaturing solution such as NaOH, Tris base, or urea.
- Precipitate the α-keratose fraction by titrating the extraction solution to a pH of 4.6 using HCl.

- Isolate the off-white α-keratose fraction by centrifugation.
- Retain the liquid fraction that contains the γ-keratose. The solution should be concentrated 10-fold by distillation of excess water using a rotary evaporator.
- Precipitate the γ-keratose by dropwise addition of the resulting solution into an eightfold excess of cold ethanol.
- Recover the off-white solid by centrifugation.
- Dissolve the α-keratose (white solid) in 20 mM Tris base with 20 mM EDTA.
- Remove any undissolved material by filtration/centrifugation.
- Reprecipitate the α-keratose by reducing the pH with slow addition of hydrochloric acid to a final pH of 4.6 (three precipitations in total).
- Neutralise the solution (pH 7.0) and remove any precipitate that forms by centrifugation/filtration.
- Dialyse using tubing with an MWCO of ca. 14.2 kDa against DI water.
- After dialysis, isolate the α-keratose again by precipitation at pH 4.6, separate by centrifugation/filtration, and freeze-dry.
- Dissolve the γ-keratose in a minimum amount of 20 mM Tris base with 20 mM EDTA.
- Reduce the pH with slow addition of HCl to a final PH value of 4.6.
- Remove any precipitate that forms by centrifugation/filtration.
- Precipitate the γ-keratose by dropwise addition to an eightfold excess of cold ethanol.
- Repeat the reprecipitation using a minimum amount of 20 mM Tris base with 20 mM EDTA (three precipitations in total).
- Neutralise the solution (pH 7.0) and remove any precipitate that forms by filtration/centrifugation.
- Dialyse using tubing with an MWCO of ca. 3.5 kDa against DI water (Fig. M.2).
- Isolate γ-keratose again by titrating to pH 4.6, removing any solids that form by filtration and discarding, and precipitating by dropwise addition to an eightfold excess of cold ethanol.
- Separate the γ-keratose by filtration/centrifugation and lyophilise.

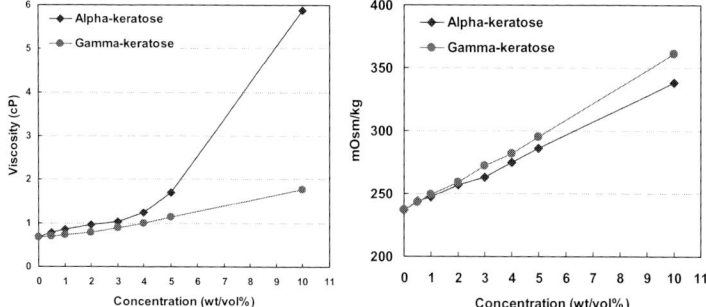

Fig. M.2 Viscosity and osmolarity of α-keratose and γ-keratose solutions with different concentrations at 37°C.

M.4 Isolation of Kerateines

- Extract kerateines by a reduction method.
- Prepare a 1M solution of TGA in DI water.
- Adjust the pH of the solution to 10.2 by the addition of saturated NaOH solution.
- Carry out the extraction in a plastic container of sufficient volume to permit vigorous mixing. Seal the container and shake at 4°C for 24 h (or 37°C for 12 h), followed by equilibration at RT.
- Precipitate the α-kerateine fraction by titrating the reduction solution to a pH of 4.6 using concentrated HCl.
- Isolate the off-white α-kerateine fraction by centrifugation/filtration.
- Retain the liquid fraction that contains the γ-kerateine.
- Concentrate the γ-kerateine solution by 10-fold on a rotary evaporator at a bath temperature of not more than 50°C.
- Precipitate the γ-kerateine by dropwise addition of the solution into an 8-fold excess of cold ethanol.
- Isolate the off-white solid by centrifugation/filtration.
- Dissolve the α-kerateine (white solid) in 20 mM Tris base with 20 mM EDTA.
- Remove any undissolved material by centrifugation/filtration.
- Reprecipitate the α-kerateine by reducing the pH with slow addition of HCl to a final pH of 4.6 (three precipitations in total).

- Neutralise the solution (pH 7.0) and remove any precipitate that forms by centrifugation/filtration.
- Dialyse using tubing with an MWCO of ca. 14.2 kDa against DI water.
- After dialysis, isolate the α-kerateine again by precipitation at pH 4.6, separate by centrifugation/filtration, and freeze-dry.
- Dissolve the γ-kerateine in a minimum amount of 20 mM Tris base with 20 mM EDTA.
- Reduce the pH with slow addition of HCl to a final PH value of 4.6. Remove and discard any precipitate that forms by centrifugation/filtration.
- Precipitate the γ-kerateine by dropwise addition of an eightfold excess of cold ethanol (three precipitations in total).
- Neutralise the solution (pH 7.0) and remove any precipitate that forms by centrifugation/filtration.
- Dialyse using tubing with an MWCO of ca. 3.5 kDa against DI water (Fig. M.3).
- Isolate the γ-kerateine by titrating to pH 4.6, removing any solids that form by filtration/centrifugation and discarding, and precipitating by dropwise addition of an eightfold excess of cold ethanol.
- Separate the γ-kerateine by centrifugation/filtration and freeze-dry (Fig. M.4).

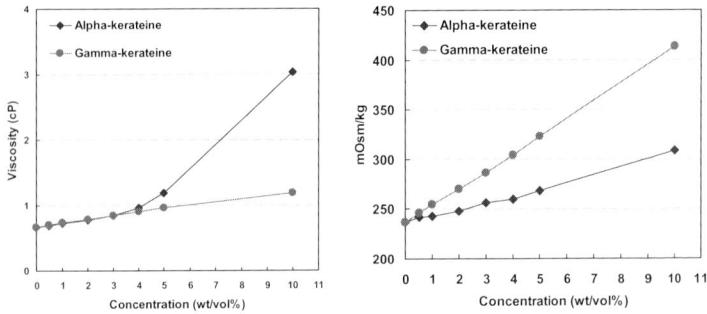

Fig. M.3 Viscosity and osmolarity of α-kerateine and γ-kerateine solutions with different concentrations at 37°C.

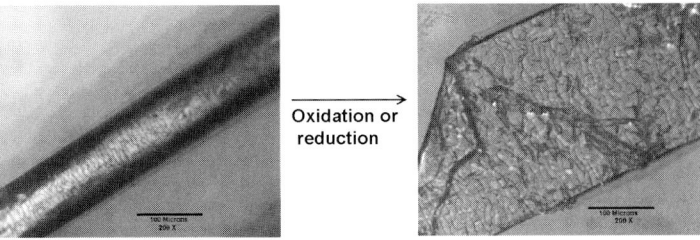

Fig. M.4 Oxidation or reduction, followed by extraction in aqueous base, removes essentially all of the cortical proteins from hair fibres.

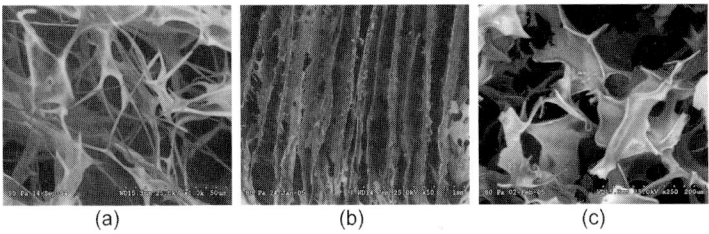

(a)   (b)   (c)

Fig. M.5 Keratin biomaterial scaffolds are formed spontaneously by a self-assembly mechanism. Keratin biomaterials can be used to build up a variety of microstructures by controlling their composition and processing.

M.5 Fabrication of a Keratin Scaffold Using Self-Assembly

- Keratin biomaterials are dissolved in DI water to make 0.5%–10% (w/v) solutions.
- The keratin solutions are poured into a plastic mould and are subsequently frozen at −20°C or −80°C for 24 h.
- After freezing, the water is removed by lyophilisation over a period of 2–3 days, depending on the conditions and sample volume.
- The fabricated keratin scaffolds are stored in a vacuum container until use (Fig. M.5).

M.6 Requirements

1. Human hair
2. Organic solvents
3. Chemical reagents
4. Deionised water
5. Shaker
6. Sieve
7. Filter system
8. pH meter
9. Centrifuge
10. Rotary evaporator
11. Dialysis system
12. Freezer
13. Lyophiliser

M.7 Cautions

- Both α-kerateine and γ-kerateine samples are prone to recross-linking.
- If dissolution in 20 mM Tris solution is unusually difficult, then cross-linking of the cysteine residues has likely occurred.
- To avoid this, kerateine samples should be stored under nitrogen at subfreezing temperature.
- If cross-linking problems persist, chemical methods can be employed to cap the sulfhydryl groups.

N

Protocol for Nanofibre Electrospinning Scaffold

Jung Keug Park

N.1 Concept

- Nanofibres — fibres with a diameter of some nm to 1 μm — are being widely studied in the field of fibre-related technology.
- Nonwoven fabrics made of nanofibres — which have soft properties and a high surface area — can be used not only as separation materials, but also as composite materials. They have high miscibility with other materials due to high fibre density (fibres/unit area) and large space between nanofibres.
- Nanofibres are very soft like skin, are thinner than paper, easily evaporate water, and do not permit the invasion of foreign materials such as bacteria from outside.
- It is possible to make bioabsorbable bandages or artificial skins made of artificial protein nanofibres that are similar to natural proteins or tissues.
- General fibres are made by extruding raw materials through a conduit with a diameter of 0.12–0.2 mm under high pressure, while nanofibres are made by dissolution spinning or electrospinning.
- Dissolution spinning, in which polymers are plasticised and additives like drugs are decomposed, is inappropriate to produce medical products. However, electrospinning, which obtains polymer jet by applying a high electric charge, can be used to produce medical products.
- Electrospun nanofibres are structurally very similar to ECM, so they can improve cell attachment and proliferation.
- The electrospinning principle, developed in the 1930s, is that electrical repulsion between raw polymeric materials induces molecular aggregation followed by division into nanosized fibres, and an unstable ejection having a similar movement as whip (narrow flow of solution and nanoparticles) through air when the raw polymeric materials are placed in a high electrical field by high voltage.
- Electrospinning is a method to obtain nonwoven fabrics as nanosized polymeric fibres from the end of a syringe are collected on a grounded collection drum charged oppositely to the syringe.
- Nanofibres with a diameter of 10–1000 nm are ejected in the whipping movement during the evaporation of solvent. Just collecting these nanofibres makes nonwoven intertwined fabrics without knitting.

- A scientific basis of electrospinning developed by Raleigh in 1882 states that electric force could overcome surface tension at liquid dropping, but electrospun fibre has a disadvantage of low mechanical properties.

N.2 Procedure

- Dissolve collagen (bovine type I collagen, Bioland Ltd., Korea) into HFP (Fluka, Germany) at a concentration of 10% [Fig. N.1(a)].
- Wrap aluminum foil on the grounded collection drum [Fig. N.1(b)].
- Pour the solution into a syringe [Fig. N.1(c)].
- Connect a conducting wire to the syringe needle [Fig. N.1(d)].
- Switch on the main power [Fig. N.1(e)].
- Set the parameters: rotation velocity of drum, voltage, and current (voltage: 14.1 kV; current: 0.01 mA) [Fig. N.1(f)].
- Switch on the output power [Fig. N.1(g)].

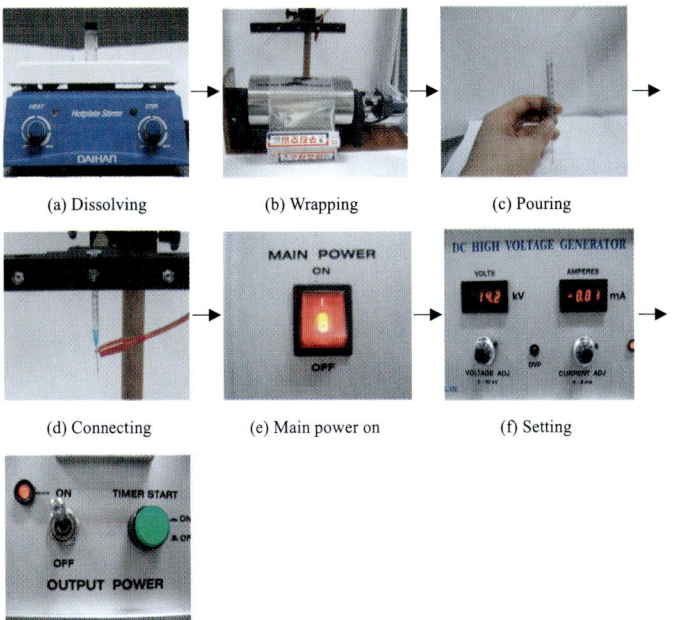

(a) Dissolving   (b) Wrapping   (c) Pouring

(d) Connecting   (e) Main power on   (f) Setting

(g) Output power on

Fig. N.1 Schematic diagram for the preparation processing of scaffolds by electrospinning.

N.3 Requirements

1. Polymer
2. Solvent
3. Magnetic bar
4. Magnetic stirrer
5. Syringe
6. Foil
7. Electrospinning system

Figure N.2 shows an electrospinning system (left) and control box (right).

Fig. N.2 Photograph of electrospinning system.

Figure N.3 shows a fabric made by collecting electrospun nanofibres, and Fig. N.4 shows nanofibres being electrospun from the needle of a syringe.

Fig. N.3 After electrospinning.

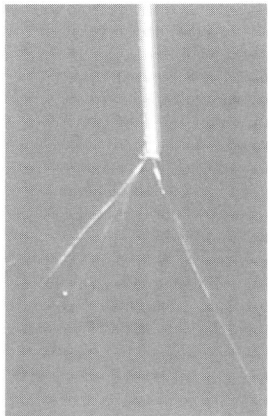

Fig. N.4 During electrospinning.

Figure N.5 shows a schematic diagram of an electrospinning system.

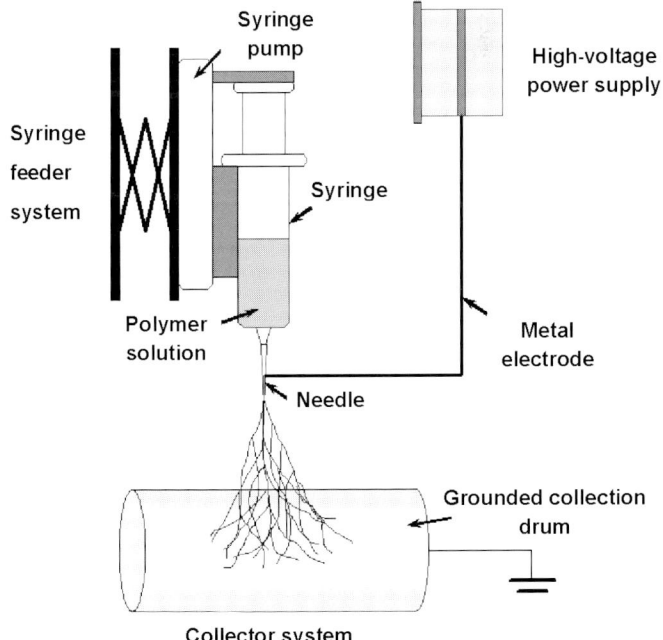

Fig. N.5 Schematic diagram of electrospinning.

N.4 Characterisations

- Figure N.6 shows correlation graphs between spinning parameters and the diameter of nanofibre. As the concentration and viscosity of solution, ejection velocity, and relative humidity increase, the diameter of nanofibre increases.
- As the current, distance between jet and drum, and salt concentration increase, the diameter of nanofibre decreases. The correlation between spinning parameters and diameter is investigated under the same conditions as those of the others.

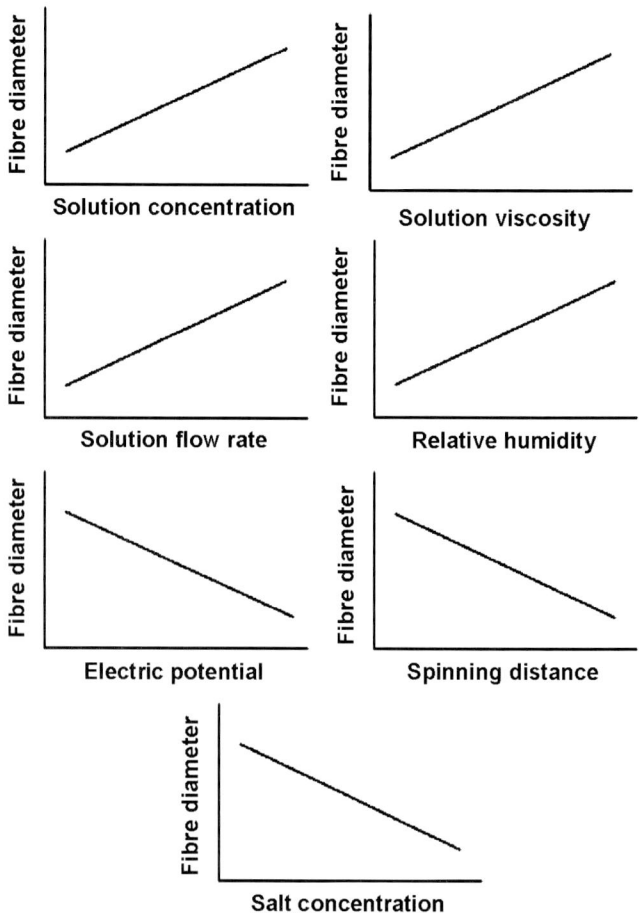

Fig. N.6 Effects of spinning parameters.

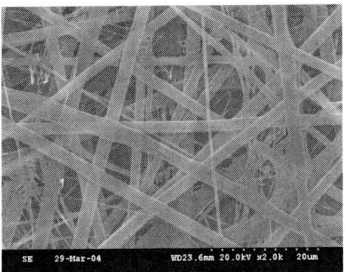

Fig. N.7 Electrospun collagen scaffold.

- Figure N.7 shows electrospun collagen fabrics by 10% collagen in HFP. The left image is an SEM of the fabrics' surface, and the right image is an SEM of their cross-section.

N.5 Cautions

- The type of solute and solvent as well as the concentration and viscosity of solutions affect the fluid dynamics. Nanofibres can be made at the roving of solution under high voltage. It is known that the available viscosity is 0.5–50 poises to make nanofibres.
- Increasing the diameter of the syringe needle reduces the diameter of nanofibre to 50 nm. Decreasing the distance between needle and drum inhibits fibre extension, so thick fibres are obtained.
- The spinning voltage is generally 10–50 kV. As the electric charge in solution increases, the diameter of nanofibre decreases. The ionic quantity generated by salts affects the force of charge at the jet. The addition of salts increases charge density at the surface of the jet during electrospinning. So, electric force is increased at the jet, bead formation is reduced, and more fine fibres are obtained.

O

Preparation of Microfibres and Fibrous Scaffolds by Wet Spinning Method

Seung Jin Lee

O.1 Concept

- Fibre spinning may be broadly divided into three categories: melt spinning method, wet solution spinning method, and electrospinning method.
- In melt spinning, the spray of molten polymer is cooled to form solid threads. Electrospinning is an effective process for forming nanofibres through a high electric field. Wet spinning is based on the extrusion of concentrated polymer solutions under pressure into precipitation solution.
- In wet spinning, the polymer solution is extruded into a nonsolvent to precipitate the polymer in the form of a thread. This technique avoids the large capital, space, and raw material requirements of the conventional melt extrusion of polymers.
- Furthermore, the fibre properties are controllable for specific applications and are able to compose synthetic and natural polymers, resulting in biocompatible fibres with good mechanical properties.
- This technique may also have the potential to incorporate heat-sensitive drugs into the fibre, as the entire process takes place at room temperature.

O.2 Procedure

There are two different wet spinning methods by grouping the physical properties of polymers and extrusion condition. The following is a description of the concrete wet spinning method of both polyester and chitosan monofilaments on half of wet spun fibres.

O.2.1 Polyester monofilament fibres

1. Dissolution

    - PLGA, PLA, and PCL are dissolved in a variety of solvent systems (including methylene chloride, DMF, chloroform, acetone, HFP, and their mixture solutions) at various concentrations from 5 to 20 wt/v%.

Note: Solution spinning of the polymers normally requires high solution viscosities to enable extrusion of a filament prior to drawing. The wet spinning technique is dependent on two key factors: the viscosity of the polymer solution, and the choice of solvent/nonsolvent systems.

2. Extrusion

- The polymer solution is loaded into a glass syringe and placed in a syringe pump, and then the syringe is connected to a 21–25-gauge needle (Fig. O.1).
- Only blunt-tipped needles should be used because the slope on sharp needles would cause problems during extrusion. The extrusion flow rate is 0.05–0.20 mL/min.
- The fibre production rate tends to increase at lower solution concentrations in line with the higher follow rate of lower viscosity solutions through a spinneret.

3. Coagulation

- The needle tip is immersed in a small stainless container full of coagulating bath fluid, which is a poor solvent for the polymer and includes isopropyl alcohol, methanol, and short alkanes.

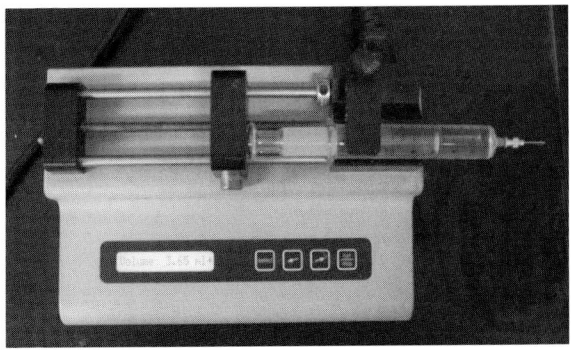

Fig. O.1 Extrusion system.

- The coagulation solution is a poor solvent for the polymer, yet highly miscible with the solvent of dissolving the polymer.
- The morphological characteristics of the filament, such as the presence of macropores and noncircular cross-sectional shapes, are greatly influenced by both temperature and the solvent content in the coagulation bath.

4. Collection

- A scheme of wet spinning line used to produce regenerated polyester fibres is reported in Fig. O.2.

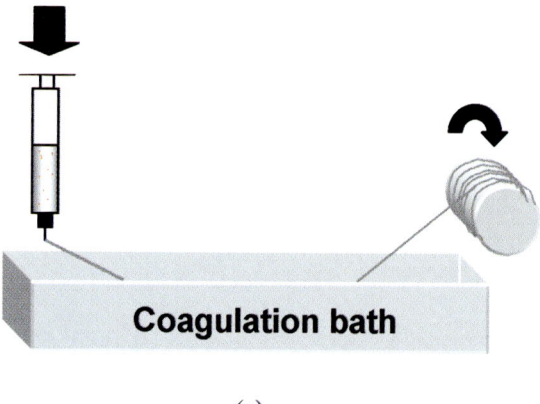

(a)

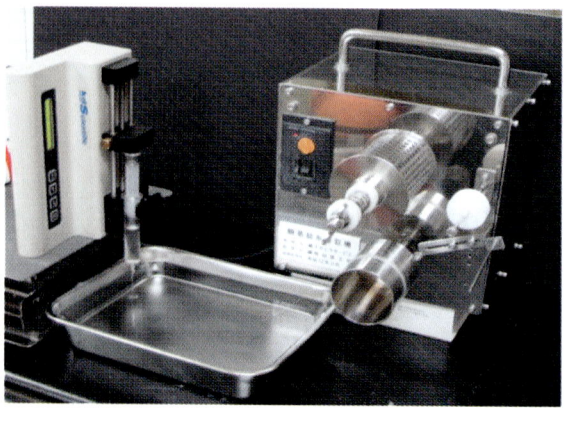

(b)

Fig. O.2 (a) Scheme and (b) photograph of wet spinning line.

- This spinning line is composed of an extrusion unit, coagulation bath, take-up, and roller.
- The collecting rate can be varied so that fibres are collected with different drawn ratios. Collected fibres are dried under vacuum for 48 h.

5. Bonding

- Fibrous matrices are shaped to have a pore size of 150–300 μm. The fibrous matrices are exposed to organic solvent vapour for 3–12 h to bond the fibres.

O.2.2 Chitosan monofilament fibre

1. Dissolution

- The chitosan powder is dissolved in aq. 2%–5% (v/v) acetic acid to produce 5%–15% chitosan dope solution, and then the solution is filtered by cloth fabric to remove insoluble particles and finally degassed in ultrasonic bath.

2. Extrusion

- Multifilament spinning of the chitosan solution is carried out with a small and simple wet spinning system, as shown in Fig. O.3. The chitosan solution is introduced into the cylinder.
- The polymer solution is extruded through a spinneret (60 holes, 100-μm diameter) submersed in a coagulation

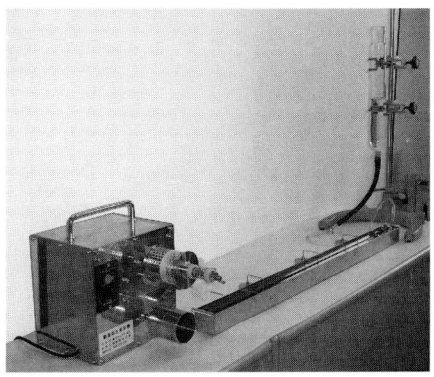

Fig. O.3 Photograph of multifilament spinning system.

Fig. O.4 Drawn fibre.

Fig. O.5 Wet spun chitosan fibre bundle.

bath containing alkali ethanol solution. A pressure of 0.2 MPa is applied.

3. Coagulation and collection

- The solution is injected into a coagulation bath containing a mixture of 1M NaOH, 50% ethylene glycol, and 50% distilled water.
- The concentration, composition, and temperature of the coagulation solution are important process variables. In the bath, the fibre bundle is drawn to get molecular chain orientation and improve the mechanical properties (Figs. O.4–O.6).
- The spools with fibres are washed in running water for 3 days, and then dried by ethanol substitution at room temperature.

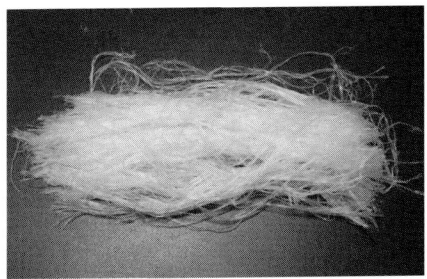

Fig. O.6 Dry chitosan fibres.

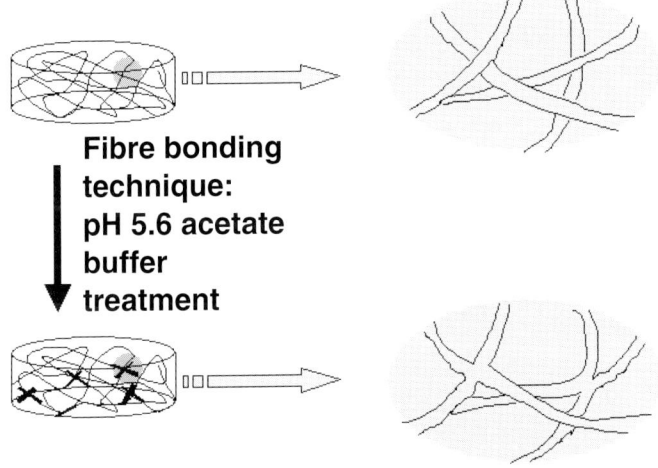

Fig. O.7 Scheme of the chitosan fibrous matrix system using fibre bonding technique by acid treatment.

4. Fibrous matrix preparation

- Three-dimensional scaffolds are developed by an acid-treated fibre bonding technique (Fig. O.7). Chitosan disks are fabricated by pressing dried chitosan fibres.
- The chitosan disks are equilibrated in water for pore generation. The matrices are subjected to a pH 5.6 acetate buffer for 15 s for partially solubilising chitosan fibres.
- The matrices are quickly placed in a 1N NaOH solution for inhibiting solubilising polymer fibres. Nonwoven chitosan fibrous matrix is washed in distilled water and freeze-dried (Fig. O.8).

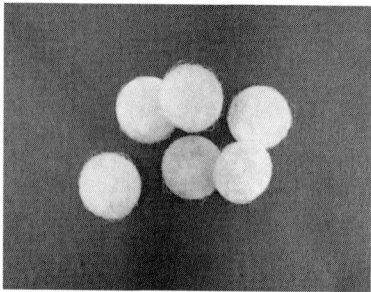

Fig. O.8 Photograph of chitosan fibrous matrix.

O.3 Characterisations

- The fibre mechanical, thermal, and rheological solution properties are characterised.
- The shear viscosity of the polymer solution is monitored using a rheometric expansion system.
- The viscoelasticity of the polymer solution, depending on the concentration and the polymer solvent interaction, is considered as one of the key factors of spinnability and physical properties of monofilaments.
- DSC analysis is performed using a thermal analysis instrument.
- To investigate the crystalline structure and molecular orientation of the regenerated polymer fibres, X-ray scattering is measured.
- To evaluate the mechanical properties of filament, stress–strain curves are obtained using Instron.
- The cross-section and surfaces of monofilaments are examined using SEM after gold coating.

P

Fabrication of Nonwoven Fibre by Melt-Blown Process

Myung Seob Khil and Hak Yong Kim

P.1 Concept

- Melt-blown technology has been introduced to make micro-fibrous nonwoven mats that are effective in regenerating functional tissues or organs (ranging from bioartificial skin to functional urinary bladder and blood vessels) using cell-scaffold–based approaches.
- In this approach, a highly porous scaffold is needed to accommodate cells and to guide their growth and tissue regeneration in three dimensions.
- The fibrillar structure is important for cell attachment, proliferation, and differentiated function in tissue culturing.
- The most important function of the microstructure of a porous scaffold matrix is to provide structural cues that guide tissue development by organising cells into a specific three-dimensional architecture, and by appropriately balancing the proliferation and differentiation of cells in the 3D space.
- The nonwoven fibres in the melt-blown process are held together by a combination of entanglement and cohesive sticking.
- The biodegradeable polyesters, PGA and PDO, are commonly used for preparing scaffolds using this technique.
- Typical air temperatures range from 230°C to 360°C.

P.2 Procedure

- The polymer is dried over 24 h at a predetermined temperature under vacuum to remove the water, as shown in Fig. P.1.
- The predried polymer is fed into a melt-blown machine through a hopper.
- The polymer is heated up to the point where the molten material can be extruded through a spinning die (Fig. P.2).
- High-velocity heated air is injected near the die tip and attenuates the filaments to a finer diameter.
- The filaments are quenched with cool air, and then are blown and collected on a moving collector screen.

Fig. P.1 Drying of PGA.

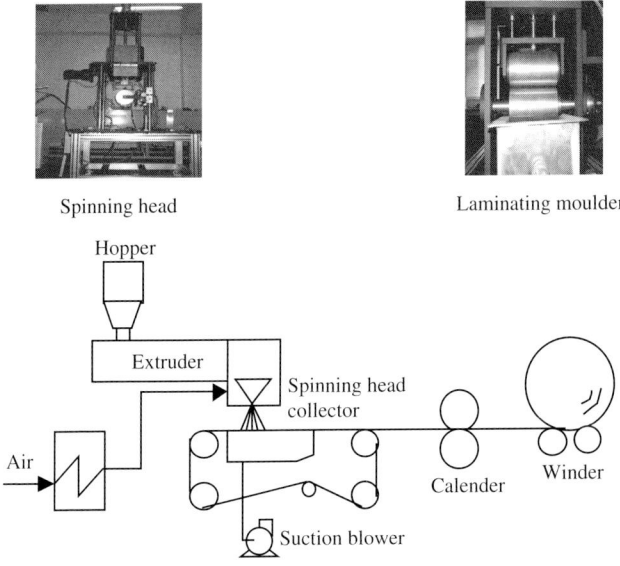

Spinning head

Laminating moulder

Fig. P.2 Schematic diagram of the melt-blown process.

- The nonwoven fibre is usually wound on a cardboard core.
- The nonwoven fibre is subsequently thermally calendered with a smooth or patterned finish, or is thermally laminated with other substrates.
- Samples are dried by heating them at a predetermined temperature under vacuum. This removes the moisture bound to the surface of the samples.

P.3 Requirements

1. Biodegradeable polymer (PGA, PDO)
2. Melt-blown apparatus
3. Calender
4. Laminator
5. Vacuum oven

P.4 Characterisations

- The diameter and polymer morphology of the melt-blown non-woven fibre are determined with the use of SEM (Fig. P.3).
- An image analyser is used to measure the distribution of the diameter.
- The following are some of the main characteristics and properties of a melt-blown nonwoven fibre:

  - The fibre orientation is random.
  - Melt-blown nonwoven fibres derive their strength from mechanical entanglement and frictional force.
  - Most melt-blown nonwoven fibres have a layered structure, with the number of layers increasing as the basis weight increases.
  - The fibre diameter ranges from 2 to 7 μm.
  - Microfibres provide a high surface area.
  - The fibres show thermal branching, as shown in Fig. P.3.

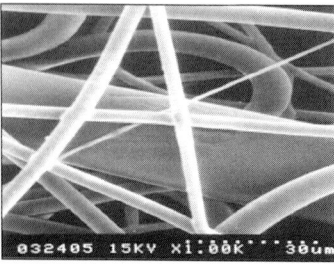
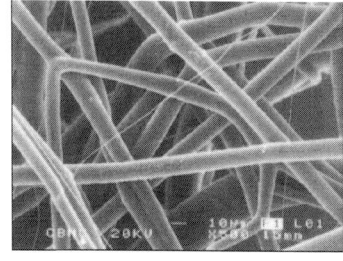

(a)   (b)

Fig. P.3 SEM images of nonwoven fibre prepared from (a) PGA and (b) PDO via melt-blown process.

P.5 Cautions

- The polymer and air throughput rates decide the final fibre diameter, the fibre entanglements, and the extent of the zone of attenuation. Therefore, they should be changed according to requirements during production.
- Bear in mind that the die and air temperatures affect the appearance and tactile hand of the fabric as well as fabric uniformity.
- The die-to-collector distance generally affects the openness of the fabric and the fibre-to-fibre thermal bonding.
- The air angle controls the nature of air flow, thus affecting the degree of fibre dispersion or turbulence.

Q

Three-Dimensional Tissue Printing Technology

Tao Xu,
Sang Jin Lee and
James J. Yoo

Q.1 Concept

- Inkjet printing technology has been developed as a tool to configure scaffolds for tissue engineering applications. Recently, efforts toward printing cells directly onto the substrates have been demonstrated with single cell types.

- To build complex tissues composed of multiple cell types and tissue components, printing a three-dimensional structure consisting of various cell types is necessary.

- Three-dimensional printing is an approach to fabricate scaffolds, wherein a binder solution is deposited onto a biomaterial powder bed layer by layer using an inkjet printer. The principle of the classic three-dimensional printing technique is shown in Fig. Q.1.

- The loading bed consists of a biomedical grade material, such as plastics and ceramics. Each layer in the build bed is incrementally raised up by a Z-axial piston as the printing progresses, depending on the type of material used and on the surface finish and accuracy needed.

- The nozzle through which the binder ejects is controlled by a computer-driven X–Y positioning system.

- The deposition of the binder in one complete layer defines the form and structure for the printed layer, and the process is repeated until all of the layers have been fabricated.

- The conventional printing technology uses toxic solvents, which are not suitable for tissue engineering applications.

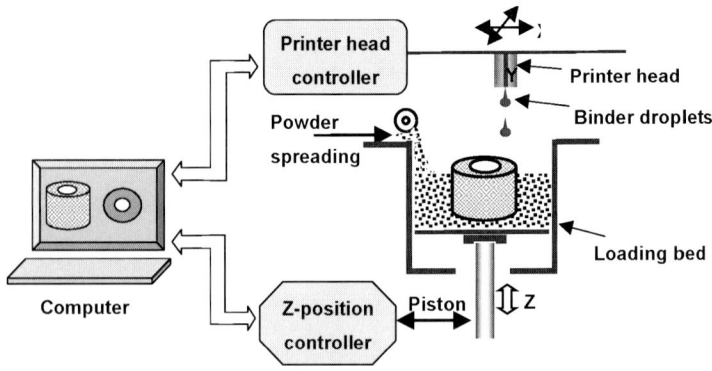

Fig. Q.1 A schematic diagram of the classic three-dimensional printing technology.

To overcome this limitation, recent developments allow the use of noncytotoxic cross-linkers and hydrogels to print three-dimensional structures.

- Noncytotoxic cross-linkers, such as $CaCl_2$, are loaded into the printer cartridges and replace the binders.
- Subsequently, the cross-linkers are ejected into the uncross-linked hydrogel, e.g. sodium alginate, which is loaded into the loading bed of the printer.
- When the cross-linkers are in contact with the uncross-linked hydrogel, rapid polymerisation occurs, resulting in rigid gelation.
- A layer-by-layer printing of the hydrogels results in a three-dimensional hydrogel scaffold with the formation of predetermined structures.
- More recently, the possibility of inkjet printing of living cells has been demonstrated. As with the printing of biomaterials, living cells can be loaded directly into the printer cartridges and printed into appropriate substrates to form cell structures.
- When combined with the printing of hydrogel, cell printing technology offers the possibility to fabricate complex 3D tissue-like constructs.

Q.2 Procedure

The desktop printer (HP 550) and cartridges are modified to permit the printing of cells and biomaterials. The paper-feeding mechanism is disabled, and a customised Z stage and chamber are installed. The printer cartridges are emptied and sterilised to accommodate cells and biomaterials.

- Sodium alginate solutions (2%) and $CaCl_2$ (0.25M) are prepared and sterilised.
- Muscle cell pallets are prepared by trypsinisation and centrifugation.
- The cells and $CaCl_2$ are mixed and loaded into the printer cartridges.
- The print designs are created using Microsoft PowerPoint® software (Fig. Q.2).
- $CaCl_2$ with cells are printed into the uncross-linked alginate gel loaded onto the Z stage and chamber.

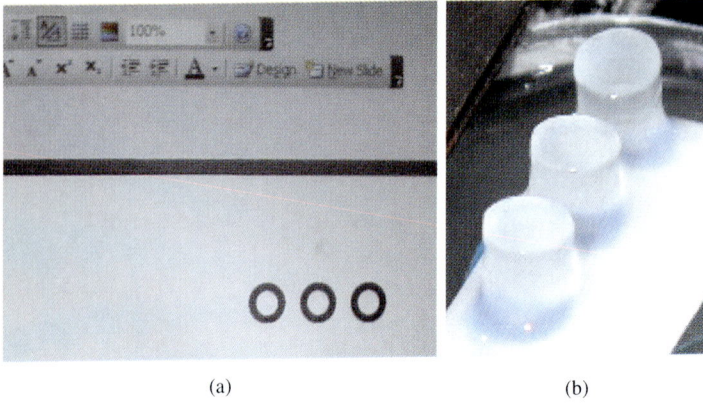

(a)   (b)

Fig. Q.2 Fabrication of three-dimensional tubular structures. (a) Ring patterns are designed with the Microsoft Powerpoint® software. (b) The tubular structures, consisting of muscle cells and alginate gels, are fabricated by layer-by-layer printing.

- After every print action, an elevator platform is lowered to refill uncross-linked alginate gel onto the Z stage for the next layer printing.
- A layer-by-layer printing of alginate gels mixed with cells results in a three-dimensional structure.

Q.3 Characterisations

- To demonstrate that the printed cells retain their phenotypic and functional expressions, a series of assays is performed.
- The viability of cell-printed three-dimensional constructs is evaluated by a two-colour fluorescence live/dead assay using calcein AM and ethidium homodimer (EthD-1).
- The live/dead experiments show that more than 90% of cells survive during the nozzle firing [Fig. Q.3(a)].
- The proliferative ability of cell-printed constructs is analysed using the mitochondrial metabolic (MTT) activity assay.
- The MTT proliferation assay shows that smooth muscle cells within the printed constructs gradually proliferate during a 7-day period [Fig. Q.3(b)].

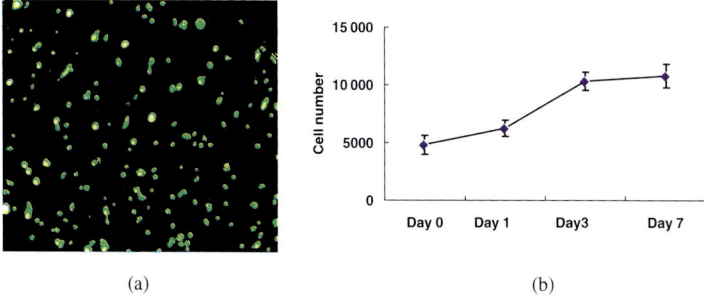

(a)   (b)

Fig. Q.3 (a) Representative live/dead fluorescent image of smooth muscle cells within the cell-printed 3D constructs (day 4). Living cells after printing are stained in fluorescent green, and dead cells in fluorescent red. (b) MTT cell proliferation assay of smooth muscle cells within 3D cell-printed constructs.

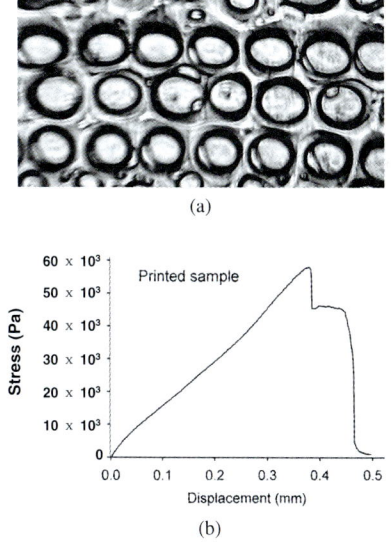

(a)

(b)

Fig. Q.4 (a) Light microscopy of the printed scaffolds. (b) Stress–displacement profiles of the printed three-dimensional constructs.

- The microarchitecture of the printed scaffolds is evaluated by light microscopy and SEM.
- The three-dimensional printed scaffolds exhibit a unique microarchitecture, consisting of numerous microshells distributed in an orderly fashion within the matrix [Fig. Q.4(a)].

- The mechanical properties of the printed scaffolds are measured at room temperature by applying the tensile stress of the sample at 5 mm/min.
- The resulting stress–strain curve is used to calculate the linear modulus and ultimate tensile stress [Fig. Q.4(b)].

R

Protocol for Melt-Based Rapid Prototyping

Rui A. Sousa,
Paula Sol,
Nuno M. Neves and
Rui L. Reis

R.1 Concept

- Computer-aided tissue engineering (CATE) covers the integration of rapid prototyping (RP) with computer-aided design (CAD) and medical imaging acquisition/processing methodologies for the production of anatomically adapted scaffolds featuring a tailored internal architecture for tissue engineering (TE) applications.

- In the TE research field, a great deal of attention has been given to the development of scaffolds by melt-based extrusion-based RP.

- In techniques such as fused deposition modelling (FDM), a filament of material is heated up and melted in an extrusion nozzle and deposited layer by layer, according to a programmeable path.

- Variations of the FDM process have been employed for the development of TE scaffolds, such as precise extruding deposition (PED) in which the plasticisation of the melt is made by a rotating screw and 3D deposition process.

- In the so-called 3D deposition process (Fig. R.1), also referred to as 3D bioplotting, the material in the powder or granular form is heated and displaced inside a thermally controlled barrel above the glass transition temperature (for amorphous) or above the melting temperature (for semicrystalline) of the polymer.

Fig. R.1 Bioplotter™ system (Envisiontec GmbH, Germany).

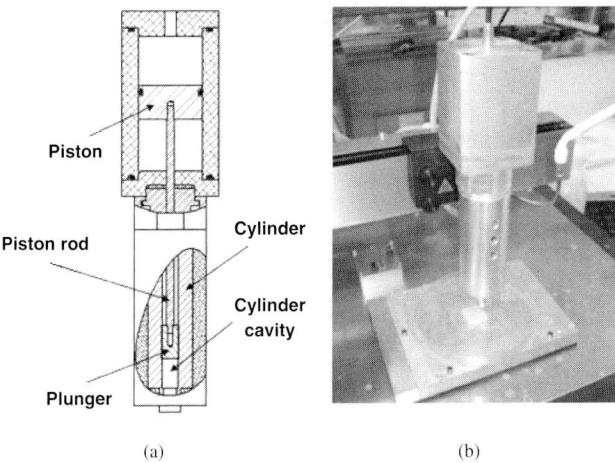

Fig. R.2 Pneumatic piston used to displace the polymer inside the barrel. (a) Schematic representation. (b) Adaptation to the Bioplotter™ equipment.

- The standard Bioplotter™ equipment employed in this chapter is modified with a pneumatic piston, which is used to displace the molten material inside the barrel (Fig. R.2).
- The following protocol is meant to produce scaffolds based on biodegradeable polymers or blends, like the blend of starch with PCL (SPCL from Novamont SpA, Italy) employed below.
- The following steps are performed using the Bioplotter™ rapid prototyping equipment (Envisiontec GmbH, Germany) and the software interface PrimCAM (PRIMUS DATA, Switzerland).

R.2 Procedure

1. Preparation of materials

   The material to be processed should ideally be in powder form. The powder form facilitates the melting of the polymer during heating as compared to granules.

   - Grind the polymer pellets using liquid nitrogen [Fig. R.3(a)]. Powder can be obtained from the polymer pellets by their respective grinding in a rotor mill employing liquid nitrogen.

(a)   (b)

(c)   (d)   (e)

Fig. R.3 Preparation of materials.

A sufficient volume of liquid nitrogen should be used in order to prevent any melting or degradation of the polymer during grinding.

- Sieve the polymer powder. Use a laboratory test sieve with a 500-μm opening. The average particle of the obtained powder should be less than 500 μm.

2. Feeding of materials

- Feed the cylinder cavity with the polymer powder or with precompacted polymer cartridges [Fig. R.3(b)]. Polymer cartridges can be prepared by cold-pressing the polymer powder into a cylindrical shape. These polymer cartridges allow for a better filling of the cavity during feeding.

3. Device configuration [Fig. R.3(c)]

- Select a luer-lock needle with an adequate diameter for the envisaged scaffold geometry. Always accurately check the real needle diameter under an optical microscope.
- Define the needle length to be used. Take into account the fact that the volume of dispensed material (V) depends on the needle length (L) and radius (r), pressure drop (ΔP), and polymer viscosity (η), according to the Hagen–Poiseuille equation:

$$V = \frac{\pi \, \Delta P \cdot \Delta t}{8 \cdot \eta \cdot L} \cdot r^4.$$

The flow rate ($\partial V/\partial t$) is inversely dependent on the needle length. For a constant flow rate, the pressure drop is proportional to the needle length.

- Calibrate the needle length by manually positioning the needle between the two laser barriers.
- Enter the respective X-, Y-, and Z-position values into the set parameters at the PrimCAM software interface [Fig. R.3(d)].
- Define the work piece origin by manually positioning the tip of the needle at the origin position.
- Enter the respective position values into the set parameters at the PrimCAM software interface [Fig. R.3(e)]. The work piece origin defines the Z start position of the dispensing head for the dispensing process, and corresponds to the origin point of the 3D data.

4. Design of scaffold architecture

- Import an appropriate data file for the scaffolds to be produced [Fig. R.4(a)]. The PrimCAM software supports the following 3D data formats: stereolithography standard file format (STL), drawing exchange format (DXF), and common layer interface (CLI). The example presented here corresponds to a $20 \times 20 \times 20$ mm^3 cube (DXF format).
- Define the internal scaffold's architecture using the PrimCAM software interface [Fig. R.4(b)]. Scaffold building parameters include the type of pattern, strand distance, offset distance, and layer thickness.
- The schematic representation of two scaffold architectures is given here as examples [Fig. R.4(c)]. (i) is an orientation pattern between consecutive layers of 0°/90° and no offset fibre distance between consecutive layers (used here as a reference); while (ii) is an orientation pattern between consecutive layers of 0°/90° and an offset fibre distance (strand distance/2) between consecutive layers.
- Preview the scaffold architecture using the Editor option of PrimCAM [Fig. R.4(d)]. The 3D model is sliced into several layers according to the defined layer thickness.
- Verify the dispensing path for each layer [Fig. R.4(e)]. Take into account the origin position for the dispensing path.
- Generate a computer numeric control (CNC) code to be transferred to the Bioplotter™ [Fig. R.4(f)].

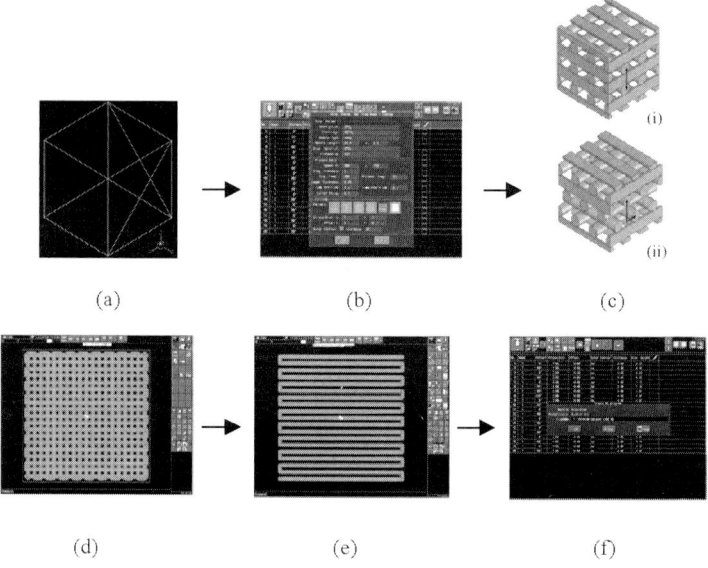

(a)   (b)   (c)

(d)   (e)   (f)

Fig. R.4 Design of scaffold architecture.

R.3 Characterisations

R.3.1 Processing conditions

- Define an adequate melting temperature [Fig. R.5(a)]. Select an appropriate melting material for processing the polymer. Always take into account the thermal stability of the polymer and the average residence time during scaffold fabrication.
- Adjust the pressure accordingly with the inherent viscosity of the polymer [Fig. R.5(b)].
- Select an adequate surface for scaffold fabrication [Fig. R.5(c)]. Select a building surface that guarantees the adhesion of the first deposited layer, taking into account the processed polymer. For the SPCL blend employed here, good results are found with cellulose filter paper substrates.
- Define the correct feeding speed [Fig. R.5(d)].

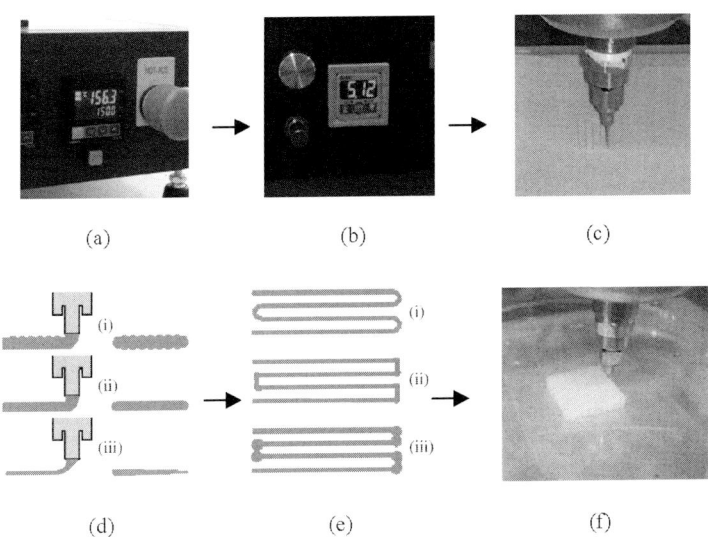

Fig. R.5 Processing condition.

- Most of the scaffold architecture is defined according to the parameters.
- Nevertheless, the feeding speed is still adjustable during layer deposition. The selection of an adequate feeding speed is important to assure constant dimension of the filaments and adhesion between layers. Low feeding speeds (i) cause the filament to be irregular due to dimensional instability of the deposited polymer, while high feeding rates (iii) cause stretching of the filament while depositing, thus compromising adhesion to the previous layer.
- Fine-tune the building parameters [Fig. R.5(e)].
- Some building parameters usually have to be optimised during scaffold fabrication, such as the corner delay time. The corner delay is the time given by the equipment to dispense corners. No sharp corners are obtained with low corner delays (i), while excess polymer is deposited at the corners for high corner delays (iii).
- These parameters are not adjustable during scaffold fabrication.
- Follow scaffold fabrication [Fig. R.5(f)].

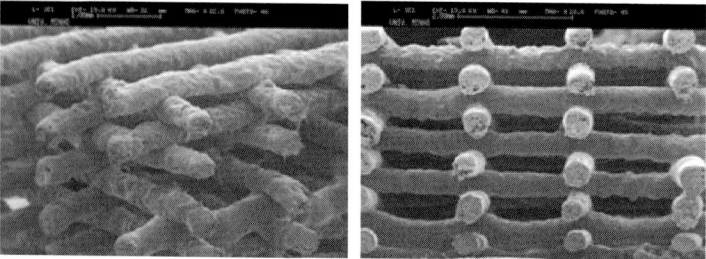

Fig. R.6 SEM of the SPCL melt-based rapid-prototyped scaffold.

R.3.2 Obtained scaffolds

- Observe the obtained scaffolds using SEM (Fig. R.6).

S

Growth Factor–Released Scaffold

Jin Woo Bae and Ki Dong Park

S.1 Concept

1. Growth factors — substances that promote the growth of cells within the body — have been extensively investigated for their regenerative potential in the tissue engineering field.

2. There are various types of growth factors, and their abilities for the control of cell-related functions are categorised by their specific therapeutic applications such as angiogenesis, bone regeneration, and wound healing, as shown below:

   - Basic fibroblast growth factor (bFGF)
   - Transforming growth factor (TGF)
   - Vascular endothelial growth factor (VEGF)
   - Platelet-derived growth factor (PDGF)
   - Bone morphogenetic protein (BMP)
   - Insulin-like growth factor (IGF)
   - Nerve growth factor (NGF)
   - Hepatocyte growth factor (HGF)
   - Etc.

3. Typically, growth factors as a signalling protein are delivered in solution form, either systemically or via direct injection into the tissue site of interest. However, growth factors have a short half-life in the body and are rapidly eliminated.

4. They may also induce adverse effects. Consequently, many researchers have studied the potential for protein delivery systems to improve the bioavailability of growth factors for an extended period.

5. There have been a number of attempts for the delivery of various growth factors using polymeric biomaterials, and these efforts have enabled an effective delivery of a specific growth factor to target tissues.

6. Growth factor–released scaffolds can be suitably designed, considering the site and the period required for applications. The control of release kinetics at a predetermined time is especially critical for the design of the delivery vehicle because growth factors are dose-dependent on exposed tissues.

7. Depending on the scaffold's geometry, porosity, volume, hydrophilicity, biodegradation, and affinity to the growth factor and site of implantation, growth factor release can either be (a) diffusion-controlled, (b) chemical and/or enzymatic

reaction–controlled, (c) solvent-controlled, or (d) controlled by a combination of these mechanisms. Materials commonly used as a scaffold for growth factor delivery are classified as natural polymers (collagen, gelatin, alginate, chitosan, fibrin) or synthetic polymers (PLGA, PLA, PEG).

S.2 Procedure

S.2.1 Physical incorporation and adsorption

Growth factors can be simply entrapped into various scaffold types such as membranes, hydrogels, foams, and particles on the fabrication process. In these steps, a whole understanding of the processing is important and the distribution of a growth factor incorporated into the matrix must be considered. Also, growth factors may be physically adsorbed on the material surface.

1. Preformed scaffold (sponge or membrane type)

   - Immerse the preformed scaffold overnight in 70% (v/v) ethanol for sterilisation, and wash with PBS solution.
   - Impregnate bFGF by immersing the scaffold in 80 μg/mL of PBS solution of bFGF for 24 h at 4°C.

2. Injectable scaffold

   - Thermosensitive hydrogel: Thermosensitive gels (PLGA-PEG-PLGA, PNIPAAm, etc.) have been widely used for biomedical applications, and the aqueous solution shows the sol–gel transition behaviour at body temperature. In this case, follow the directions as described below:

     (a) After completely dissolving 1 g of PLGA-PEG-PLGA in 4 mL of distilled water, add 10 μg of a growth factor.

     (b) After complete mixing, just use it at 37°C.

   - Microparticle: Microparticles of PLGA/PEG blends containing 0%, 1%, or 5% PEG (w/w) are fabricated using a double-emulsion, solvent-extraction technique [(water-in-oil)-in-water]. PEG is blended by modulating drug release profiles.

(a) Dissolve PLGA and PEG in a total amount of 247.5 mg in 1 mL of methylene chloride.

(b) Dissolve 1.5 µg of TGF-β_1 in 125 mL of distilled water, and inject it into a flint glass tube containing the polymer solution.

(c) Emulsify the entire mixture on a vortexer for 1 min, and then re-emulsify this solution in 100 mL of 0.3% (w/v) PVA solution.

(d) Add the second emulsion to 100 mL of a 2% IPA solution, and maintain on a magnetic stirrer for 1 h.

(e) Collect the microparticles, centrifuge, and lyophilise. (Microparticles may be embedded into 2D or 3D scaffolds.)

S.2.2 Affinity-based incorporation

1. Heparin immobilisation (Fig. S.1)

- Equilibrate EDC and NHS at room temperature.
- Incubate amine group (–NH$_2$)-containing polymers with 0.05M MES (2-morpholinoethane sulfonic acid) buffer (pH 5.6) at room temperature.
- Add EDC and NHS to heparin in 10 mL of 0.05M MES buffer at a molar ratio of EDC:NHS:heparin-COOH of 0.4:0.24:0.1.
- Activate for 10 min at room temperature, and then add it (a molar excess of preactivated heparin) to the polymer solution.
- React over 2 h at room temperature, and finally wash with distilled water for several times (or use a dialysis membrane).

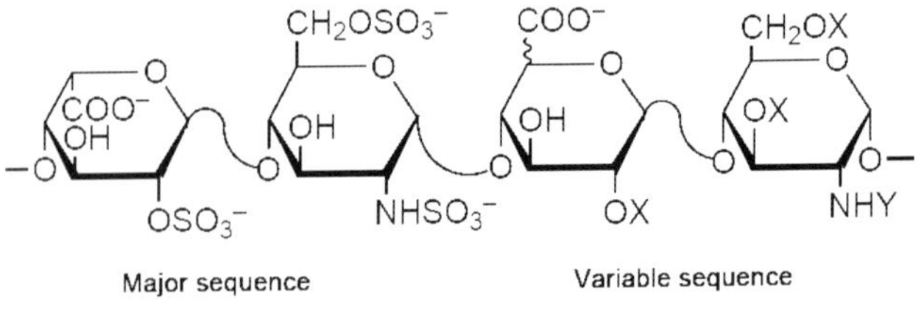

Major sequence   Variable sequence

Fig. S.1 Major and minor disaccharide repeating units in heparin (X = H or SO$_3^-$; Y = Ac, SO$_3^-$, or H).

- Incubate these samples with a heparin-binding growth factor (bFGF, TGF-β_1, PDGF, VEGF, etc.) solution for over 2 h at 37°C.

> **Note:** This coupling reaction may be carried out in other buffer conditions, not an MES buffer. However, in the cross-linking step, the ester on the NHS-activated heparin is more stable at pH 5 than at a higher pH. EDC reacts with a carboxyl group and forms an amine-reactive intermediate that is unstable in aqueous solutions. Failure to react with an amine results in hydrolysis of the intermediate, regeneration of the carboxyl, and release of an *N*-substituted urea.

2. Gelatin microspheres

- Dissolve 5 g of gelatin in 45 mL of distilled water at 60°C, and add dropwise to 250 mL of chilled olive oil while stirring at 500 rpm.
- Add 100 mL of chilled acetone (4°C) to the emulsion after 30 min. After an additional 60 min, collect the microspheres by filtration and wash with acetone.
- Cross-link the obtained microspheres in 0.1 wt% Tween 80 solution with 10 mM glutaraldehyde (GA) while stirring at 500 rpm at 15°C.
- After 15 h, collect the cross-linked microspheres by filtration, wash with water, and then agitate in a 25-mM glycine solution for 1 h to inactivate any unreacted GA.
- Collect the cross-linked microspheres by filtration, wash with water, and then lyophilise overnight. Sterilise with ethylene oxide for 16 h.
- Mix 100 mg of sterilised microspheres and 500 μL of an aqueous TGF-β_1 solution (1.2 ng TGF-β_1/mL) at pH 7.4 [at this pH, gelatin microspheres and TGF-β_1 form a complex due to the association of the negative charges of acidic gelatin (IEP of 5.0) and the positive charges of TGF-β_1 (IEP of 9.5)].
- Vortex the mixture and incubate at 4°C for 15 h.

S.3 Characterisations (Assay Procedure of a Growth Factor)

To quantitatively analyse the loaded or released growth factor, the assays of samples employ the quantitative sandwich enzyme immunoassay technique (ELISA), which is based on Quantikine® (R&D Systems, Inc., USA). The assay procedure for bFGF as a model growth factor is described below. More information related to other growth factors is available on the website of R&D Systems (www.rndsystems.com).

- bFGF

  > Bring all reagents to room temperature before use. It is recommended that all samples and standards be assayed in duplicate.

- Reagents

  1. FGF basic microplate
  2. FGF basic conjugate
  3. FGF basic standard
  4. Assay diluent RD1-43
  5. Calibrator diluent RD5-14
  6. Wash buffer concentrate
  7. Colour reagent A
  8. Colour reagent B
  9. Stop solution
  10. Plate covers

- Other supplies required

  1. Microplate reader capable of measuring absorbance at 450 nm, with the correction wavelength set at 540 nm or 570 nm (Fig. S.2).
  2. Pipettes and pipette tips

  (Continued)

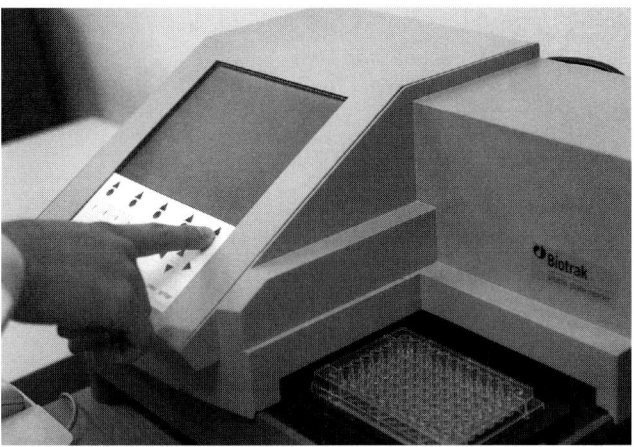

Fig. S.2 Microplate reader.

(*Continued*)

3. 500-mL graduated cylinder
4. Deionised or distilled water
5. Multichannel pipette, squirt bottle, manifold dispenser, or automated microplate washer

• Wash buffer

If crystals have formed in the concentrate, warm to room temperature and mix gently until the crystals have completely dissolved. Dilute 20 mL of wash buffer concentrate into deionised or distilled water to prepare 500 mL of wash buffer.

• Substrate solution

Colour reagents A and B should be mixed together in equal volumes within 15 min of use. Protect from light. About 200 μL of the resultant mixture is required per well.

- FGF basic standard

> Reconstitute the FGF basic standard with 2 mL of calibrator diluent RD5-14. This reconstitution produces a stock solution of 640 pg/mL. Allow the standard to sit for a minimum of 15 min with gentle agitation prior to making dilutions.

Pipette 500 µL of calibrator diluent RD5-14 into each tube. Use the stock solution to produce a dilution series (Fig. S.3). Mix each tube thoroughly before the next transfer. The undiluted FGF basic standard serves as the high standard (640 pg/mL). Calibrator diluent RD5-14 serves as the zero standard (0 pg/mL). Discard the FGF basic stock solution and dilutions after 4 h. Use a fresh standard for each assay.

- Prepare all reagents and working standards as directed above.
- Remove excess microplate strips from the plate frame, return them to the foil pouch containing the desiccant pack, and reseal.
- Add 100 µL of assay diluent RD1-43 to each well. Assay diluent RD1-43 contains a precipitate. Mix well before and during its use.
- Add 100 µL of the standard or sample per well. Cover with the adhesive strip provided. Incubate for 2 h at room temperature. Provide a plate layout to record the samples and standards assayed.

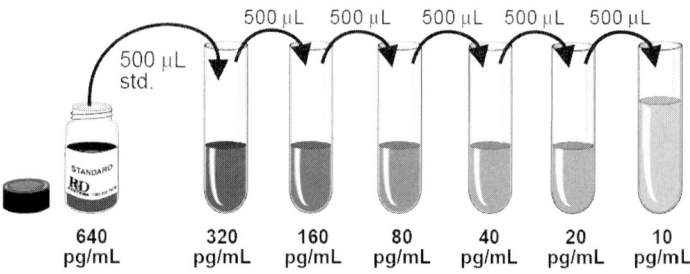

Fig. S.3 Dilution of calibrator diluent.

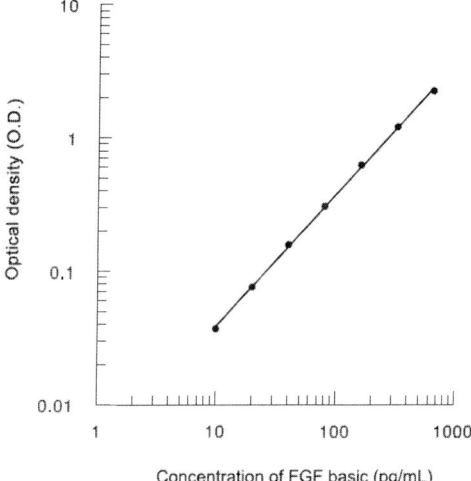

(pg/mL)	O.D.	Average	Corrected
0	0.063 0.061 0.100	0.062	-
10	0.098 0.139	0.099	0.037
20	0.136 0.223	0.138	0.076
40	0.216 0.367	0.220	0.158
80	0.366 0.688	0.366	0.304
160	0.686 1.274	0.687	0.625
320	1.264 2.310	1.269	1.207
640	2.291	2.300	2.238

Fig. S.4 Typical calibration data.

- Aspirate each well and wash, repeating the process three times for a total of four washes. Wash by filling each well with wash buffer (400 μL) using a squirt bottle, multichannel pipette, manifold dispenser, or autowasher. The complete removal of liquid at each step is essential to good performance. After the last wash, remove any remaining wash buffer by aspirating or decanting. Invert the plate and blot it against clean paper towels.

- Add 200 μL of FGF basic conjugate to each well. Cover with the adhesive strip provided. Incubate for 2 h at room temperature.
- Repeat the aspiration/wash step.
- Add 200 μL of substrate solution to each well. Incubate for 30 min at room temperature. Protect from light.
- Add 50 μL of stop solution to each well. If the colour change does not appear uniform, gently tap the plate to ensure thorough mixing.
- Determine the optical density of each well within 30 min, using a microplate reader set to 450 nm. If the wavelength correction is available, subtract the reading at 540 nm or 570 nm of the plate. Readings made directly at 450 nm without correction may be higher and less accurate (Fig. S.4).

T

Preparation of Ceramic Scaffold by Sponge Replication Method

Chang Kook You and Suk Young Kim

T.1 Concept

- The sponge replication method is designed to prepare porous 3D scaffolds for bone tissue engineering, and is a promising method for interconnected porous blocks.
- Three-dimensional interconnected porous ceramic scaffolds can be produced by the replication of porous polyurethane (PU) sponge geometry after ceramic paste coating on it and firing.
- The pore size, geometry, and porosity of ceramic scaffolds prepared by this method can be easily controlled by selecting the desired pore size of the PU sponge, ranging from 200 to 1000 µm.
- The sponge replication process involves (1) the preparation of a ceramic paste with additives, (2) the coating of the paste on sponge, (3) the drying and firing at sintering temperature with the removal of sponge, and (4) the second coating of the ceramic slurry for enhanced mechanical strength of porous ceramic scaffolds.
- Porous 3D ceramic scaffolds may support bone cell adhesion on their struts and cell ingrowth as well as blood vessel ingrowth into the pores and bone colonisation *in vivo*.
- The prepared ceramic paste has relatively high viscosity and shows viscoelastic behaviour such as dilatent rheology, where thixotropic property is essential through the coating.
- This method can be applied to any calcium phosphate ceramics. However, a nonaqueous paste formulation system is needed for degradeable calcium phosphates such as tricalcium phosphate, biphasic calcium phosphate (a mixture of hydroxyapatite and tricalcium phosphate), or calcium sulfate.
- The maximum solid loading is around 30 vol% in net volume of ceramic paste, and the desired volume of powder (which is usually dependent on the particle size) is recommended between 20 and 25 vol%.
- The formulation of liquid phase for the ceramics paste contains (1) solvents (water or alcohol), dispersants, and binders for wrapping the particles together (e.g. PVA) and giving it proper green strength; (2) plasticisers for plasticity after drying; and (3) thinners or thickeners, and drying control

chemical additives for preventing cracks on drying due to high shrinkage and fast drying.

- The ceramic powders should be nanosized by ball-milling or an equivalent manner in advance before being mixed with liquid solution.

T.2 Procedure

- The prepared PU sponges with desired shape are etched and cleaned ultrasonically in 2% NaOH solution for 15 min, and then washed in tap water and distilled water before being dried in a drying oven at 60°C [Fig. T.1(a)].
- PVA as a binder is added and stirred in distilled water at 40°C–50°C of solution temperature until the PVA becomes dissolved and fully hydrolysed without any transparent PVA gel particles [Fig. T.1(b)].
- Dispersant, plasticiser, thinner or thickener, and other additives are successively added into the PVA-dissolved water solution [Fig. T.1(c)].
- Ground ceramic powders are added to the prepared solution, which becomes a paste state (slurry) [Fig. T.1(d)].
- The paste is kneaded homogeneously; otherwise, a proper automatic kneader may be chosen for a large volume of paste [Fig. T.1(e)].
- The prepared sponges are immersed into the paste, and then squeezing operation is applied several times by compression and release in the paste, where the struts of sponges are adequately coated with the paste [Fig. T.1(f)].
- In order to remove excessive paste located in pores of sponges, the sponges are taken away from the beaker and then gently rolled with a bar or roller on a rigid plate or aluminium foil [Fig. T.1(g)].
- The paste-coated sponges are slowly dried at 30°C–35°C in air or at 60°C in a dryer oven if a drying control chemical additive is added into the paste [Fig. T.1(h)].
- After the completion of drying, the dried samples are very slowly heat-treated and sintered at each sintering temperature according to an appropriate heating schedule [Fig. T.1(i)].

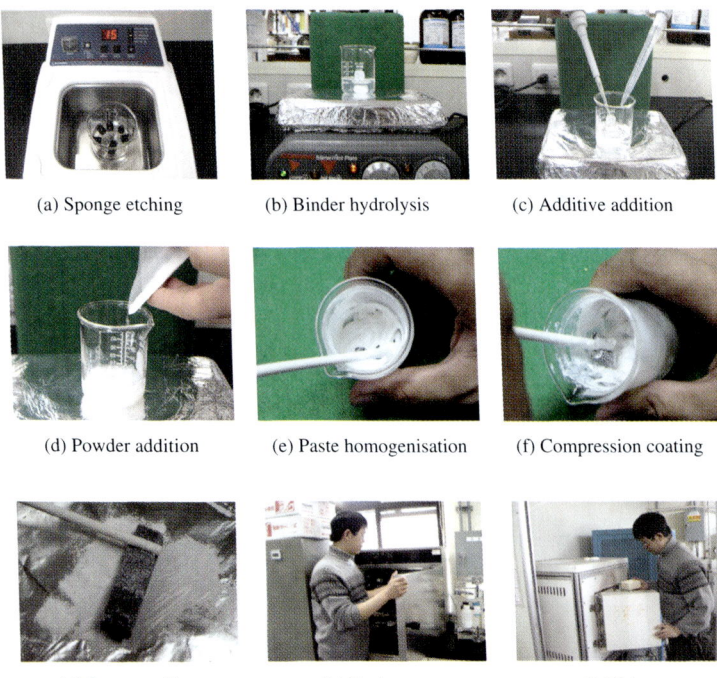

(a) Sponge etching	(b) Binder hydrolysis	(c) Additive addition
(d) Powder addition	(e) Paste homogenisation	(f) Compression coating
(g) Sponge rolling	(h) Drying	(i) Firing

Fig. T.1 Preparation processing of ceramic scaffolds by polymeric sponge method.

T.3 Requirements

1. Ultrasonic cleaner
2. Heating stirrer
3. Pipette
4. Beaker
5. Plastic rod or roller
6. Dryer oven
7. Furnace

T.4 Characterisations

- Because the viscosity of a paste is very high, the spring-back force of a sponge after compression is very important. The etching conditions (time, concentration) in NaOH solution affect the spring-back of the sponges.

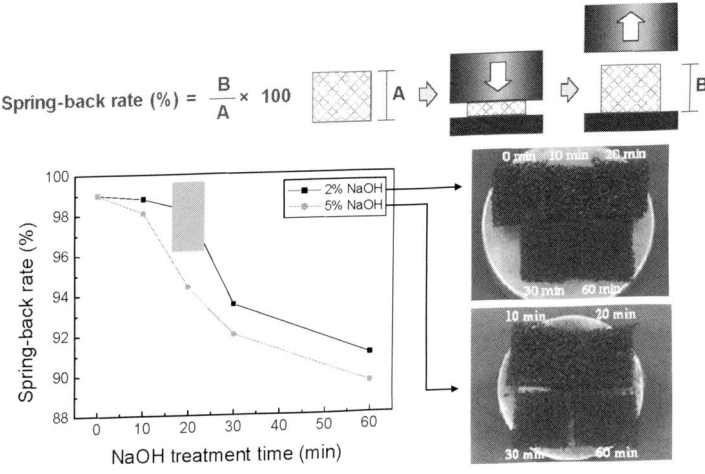

$$\text{Spring-back rate (\%)} = \frac{B}{A} \times 100$$

Fig. T.2 Spring-back rate and appearance of PU sponges after etching in 2% and 5% NaOH solution.

- After 2% NaOH treatment with different time, the spring-back rate of a sponge should be measured in terms of dimension ratio just before and after compression of a sponge, where more than 98% of spring back is allowed (Fig. T.2).
- The powder content and viscosity of the paste should be controlled differently, according to the pore size of sponges. For example, a 25/75 volume ratio between powder and solution is recommended for a 45-ppi (pores per inch) sponge (pore size distribution of 500–1000 μm), and a 20/80 ratio for a 60-ppi sponge (pore size distribution of 200–400 μm).
- The binder content can be selected based on solid loading and paste viscosity. For example, 5 wt% binder of powder weight is recommended for a 45-ppi sponge, and 3 wt% for a 60-ppi sponge (Fig. T.3).
- Various scaffold forms with various shapes and pore sizes can be easily prepared by designing the sponge shape and selecting the proper pore size of a sponge (Fig. T.4).
- Three-dimensional scaffolds by the sponge replication method show a variety of porosities up to around 95% and pore sizes ranging from 200 to 1000 μm in diameter.
- The pore size distribution can be measured by the mercury intrusion method (Table T.1), and the pore geometry and structure of struts can be examined by SEM (Fig. T.5).

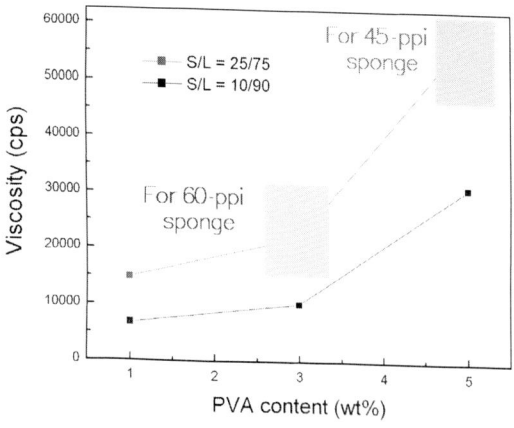

Fig. T.3 Viscosity of pastes with different PVA content as a binder and powder–liquid ratio.

Fig. T.4 Porous ceramic scaffold by a sponge method with various pore sizes.

- Apparent porosity can be measured by dividing the weight of a machined cubic scaffold over the apparent volume.
- The SEM pictures represent the microstructure replicated from sponges with an interconnected pore system, and the pore sizes of ceramic scaffolds depend on the type of sponge.

Table T.1 Pore size distribution of ceramic scaffolds prepared by sponge replication method.

Type of sponge	Pore size of sponge	Pore size after sintering
45 ppi	700–1300 µm	500–1000 µm
60 ppi	350–600 µm	200–400 µm
80 ppi	200–300 µm	100–200 µm

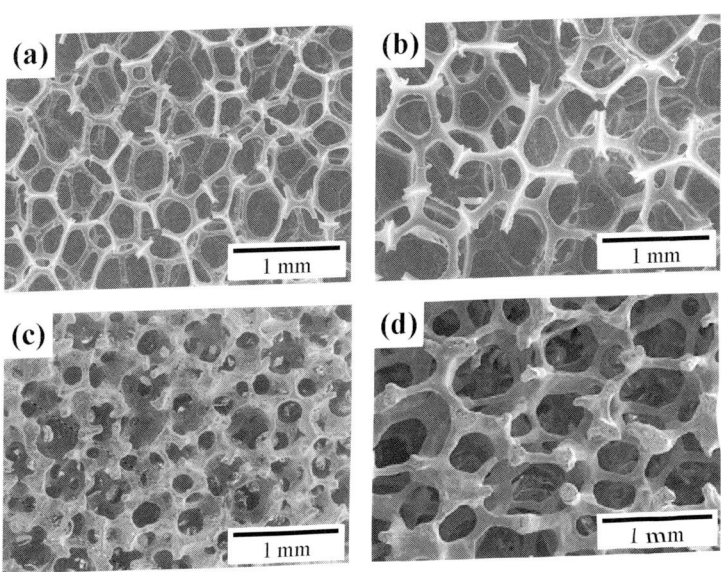

Fig. T.5 SEM pictures of (a) 60-ppi and (b) 45-ppi PU sponges, and scaffolds after sintering prepared with (c) 60-ppi and (d) 45-ppi sponges.

- The surface of struts is relatively homogeneous and dense, and the network structure may support better geometry for cell adhesion, blood vessel ingrowth, and body fluid circulation.
- Scaffolds prepared by the sponge replication method can be machined into the desired shape well, due to adaptable mechanical strength.
- If the sponges are coated with degradeable materials, it is thought that the scaffold may be degraded and substituted for newly colonised bone with a balanced time schedule.

T.5 Cautions

- Because this method is a pressureless process, the sintering property of calcium phosphates is not so excellent. There may be a limitation to obtain a proper mechanical strength.
- In order to improve the mechanical strength of as-sintered calcium phosphate scaffold, a second coating on the first-coated sintered scaffold is recommended.
- The powder content in the paste for the second coating should be reduced compared to that of the paste for the first coating. The desired powder content is recommended as 15 vol% for 45-ppi scaffolds and 10 vol% for 60-ppi scaffolds.
- Once the scaffold is immersed in the paste and taken out, it may need to be centrifuged for the removal of excessive slurry and homogeneous thin coating.
- The second drying and firing schedule is nearly the same as that of the first coating.

U

Protocol for Simple Calcium Phosphate Coating Method

Hyun-Man Kim

U.1 Concept

- Thin coating of calcium phosphate (CaP) can provide biocompatibility or enhance the bioreactivity of biomaterials for implants or scaffolds for tissue engineering without losing the physical properties inherent to the substrate materials.
- CaP nuclei attach and grow directly on the solid surfaces, including surfaces of low interfacial energy, by gradually increasing the supersaturation of the solution.
- This method can form CaP coating on most biomaterials: metals; glasses; inorganic ceramics; organic polymers including hydrophobic organic polymers such as PLGA, PS, PP, silicone, and PTFE; and organic biological tissue matrices like decalcified membranes of crab (consisting of chitin), collagens, fibres of silk, and hairs.

U.2 Procedure

- About 4.0 mM PBS, pH 7.4, is prepared by dissolving monobasic and dibasic phosphate salts in demineralised distilled water.
- Approximately 100 mL of the calcium phosphate ion solution is prepared by slowly adding 1 mL of 400 mM calcium nitrate into 99 mL of precooled phosphate buffer solution (4.0 mM PO_4, pH 7.4) at $0.0°C–3.0°C$. A peristaltic pump can be used for slow mixing if precipitation occurs.
- Sterile ion solution can be prepared by filtering the prepared ion solution using a 0.22-µm filter for direct use without extrasterilisation processing later.
- The ion solution is poured into the vessels containing biomaterials or scaffolds. Supersaturation of the CaP ion solution is gradually increased to provide a driving force for nucleation by heating the ion solution without chemical disturbance of the solution that might cause an abrupt homogeneous nucleation in the solution. The ion solution is gradually heated and kept at the same temperature in an oven set at $37°C$.

Note: The coating time is dependent on the surface energy of substrates to be coated and on the volume of the ion solution. Generally, surfaces of low interfacial energy take a long time to be coated with CaP thin film.

- The same procedure can be repeated to increase the thickness of CaP coating from nanothickness to microdimension using fresh CaP ion solution.

U.3 Characterisations

- The FTIR spectra of crystals can be obtained using a Fourier transform spectrometer, after embedding the crystals removed from the substrates in KBr pellets.
- CaP thin films formed on the surface can be observed using SEM (Fig. U.1).

U.4 Sterilisation and Uses

Seeding of cells on the CaP coating or implantation can be followed without surface extratreatment for biocompatibility after sterilisation of the substrate using 100% ethanol overnight, or directly without sterilisation if the sterilised ion solution is used.

U.5 Cautions

Rapid stirring of the ion solution is recommended so that no precipitation is formed when the solution comprising

(Continued)

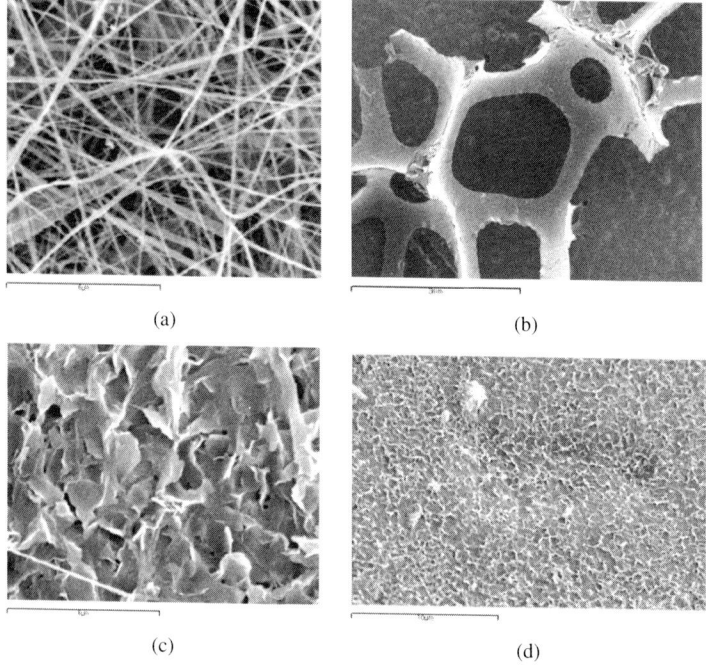

(a)

(b)

(c)

(d)

Fig. U.1 Calcium phosphate coating on (a) collagen microfibrils and (b) PU; (c) and (d) are high-magnification views of CaP thin film formed on the surface of the above viewing materials, respectively. The surface of the substrate is totally covered with calcium phosphate crystals.

(Continued)

calcium ion is added. If there is precipitation in the solution after mixing both solutions, the precipitation particles should be removed by filtration using a 0.45–0.22-μm filter.

V

Method and Techniques for Scaffold Sterilisation

*Moon Suk Kim,
Gilson Khang and
Hai Bang Lee*

V.1 Introduction

Sterility is defined as the absence of all living organisms or absolute freedom from biological contamination. Living organisms include microorganisms such as bacteria, yeasts, moulds, and viruses. The presence of even one viable living organism on a scaffold renders it nonsterile. Sterility should not be confused with cleanliness. Sterilisation is the process to inactivate and eliminate all viable living organisms and their spores. Scaffolds introduced transiently or permanently into the body of a human or an animal must be sterile to avoid subsequent infection, which could lead to serious illness or death. If the scaffolds are sterile, no living organism growth will occur; if it is nonsterile, the scaffolds will become contaminated as a result of living organism proliferation. An acute awareness of possible risks and a clear concept of sterilisation are essential to prevent serious problems. The gas (ethylene oxide, EO) and irradiation sterilisation methods are particularly noteworthy with respect to sterilisation for scaffolds.

V.2 Concept

V.2.1 EOG sterilisation

- EO is a clear liquid below its boiling point of 11°C and a gas at ambient temperature.
- EO is toxic and is considered a human carcinogen.
- The lethal effect of EO on microorganisms is mainly due to the alkylation of amine groups on nucleic acids.
- In spite of its hazardousness, the use of EO for scaffold sterilisation has many advantages, such as its efficacy even at low temperatures, high penetration ability, and compatibility with a wide range of materials.
- The EO sterilisation process consists of three components: conditioning, sterilisation, and aeration.
- The EO sterilisation process typically ranges from 2 to 16 h in duration, depending on the time required for aeration inside the sterilisation chamber.

V.2.2 Gamma radiation sterilisation

- This method of sterilisation utilises ionising radiation that involves gamma rays from a ^{60}Co (cobalt-60) isotope source.
- The gamma rays cause ionisation of key cellular components, especially nucleic acids, resulting in the death of microorganisms.
- Therefore, gamma radiation effectively kills microorganisms throughout scaffolds and their packaging with little temperature effect.
- Gamma rays are highly deep-penetrating, and the typical doses used for the sterilisation of scaffolds are readily delivered and measured.
- Gamma radiation sterilisation is by far the most popular and widespread method for scaffolds.

V.2.3 Electron beam radiation

- Electron beam radiation is a form of ionising energy that performs best when used on low-density, uniformly packaged products.
- With this method, radioactive isotopes are not involved because the electron beam is machine-generated using an accelerator.
- The accelerator is located within a concrete room to contain stray electrons; but when the accelerator is turned off, no radiation or radioactive material is present and therefore a water-filled pool is unnecessary.
- As with gamma rays, the lethality against microorganisms is related to the ionisation of key cellular components.

V.3 Procedure

V.3.1 Ethylene oxide sterilisation

- For EO sterilisation, scaffolds contained within gas-permeable packaging are hermetically sealed, as shown in Fig. V.1.
- The packages are loaded into a sterilisation chamber (Fig. V.2).

Fig. V.1 Package procedure of scaffold.

Fig. V.2 A schematic image of a typical EO steriliser with closed chamber to maintain the desired temperature and EO concentration.

- The chamber is evacuated to remove air at a rate and to a final pressure that is compatible with the product and packaging, and then moisture is introduced to attain a relative humidity generally between 60% and 80%.
- The presence of moisture is required for sterilisation efficacy with EO gas. The EO gas is then injected to a final concentration of ~600–800 mg/L.
- The steriliser is maintained at the desired gas concentration and temperature (typically 40°C–50°C) for a sufficient time.
- The chamber is re-evacuated to remove the EO, and air flushes are performed to reduce the EO level to below acceptable limits.

V.3.2 Gamma radiation sterilisation

- A schematic top view of a typical industrial ^{60}Co irradiator is shown in Fig. V.3.

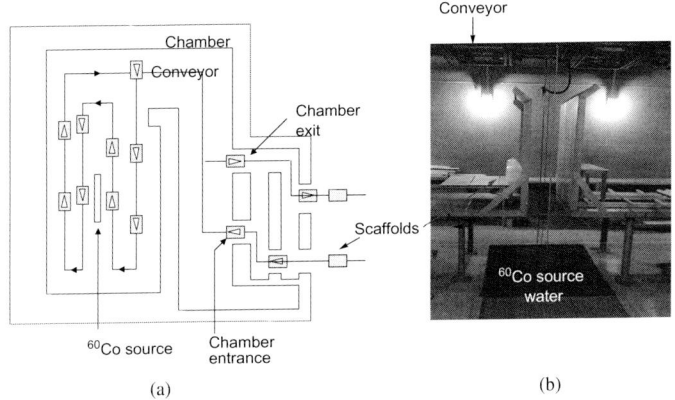

Fig. V.3 (a) A schematic image and (b) a picture of a typical ^{60}Co irradiator.

- When the irradiator is not in use, the gamma-ray source rack is lowered into a water-filled pool. When in use, the scaffolds to be sterilised are loaded into aluminium carriers or totes that are suspended from a conveyor system.
- The scaffolds are automatically conveyed into the irradiation chamber (concrete room) around the radiation source.
- The radiation source is raised to the processing position.
- The desired dose is uniformly delivered by the raised radiation source rack.
- Radiation measuring devices called dosimeters are placed along with the scaffolds to be sterilised, and the minimum and maximum doses are monitored.
- The scaffolds are brought to the unloading area in a sequential flow.

V.4 Cautions

V.4.1 EOG sterilisation

- Contact of liquid EO with the skin and eyes as well as inhalation of the gas should be avoided.

(Continued)

(Continued)

- Because of potential toxicity/carcinogenicity, residual EO and its byproduct, ethylene chlorohydrin (EC), are of concern in packaging scaffolds.
- The maximum allowable limits for EO and EC are no longer expressed as a few hundred parts per million (ppm) in scaffolds.
- The EO sterilisation process, if carried out for a long time at high temperature, may induce deformation and/or degradation of scaffolds prepared by polymers (Figs. V.4 and V.5).

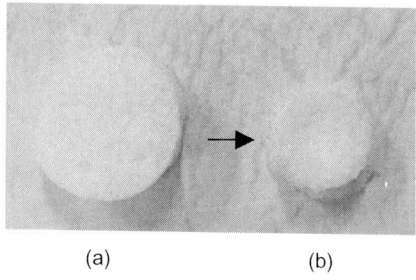

(a)   (b)

Fig. V.4 Photoimage (a) before and (b) after EO sterilisation of PLGA scaffold prepared by salt-leaching method. The scaffold shape contracts after EO sterilisation at 55°C for 4 h.

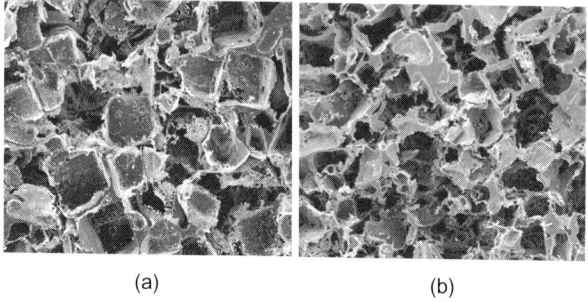

(a)   (b)

Fig. V.5 SEM microimage in cross-section (a) before and (b) after EO sterilisation of PLGA scaffold prepared by salt-leaching method (magnification is 500). The scaffold changes from an open pore structure to a closed one after EO sterilisation at 55°C for 4 h.

V.4.2 Gamma radiation sterilisation

- Some materials, e.g. PLGA, are degraded by gamma irradiation due to molecular-chain scission (Figs. V.6 and V.7).
- The fluoropolymer PTFE is not compatible with this sterilisation method because of its extreme radiation sensitivity.

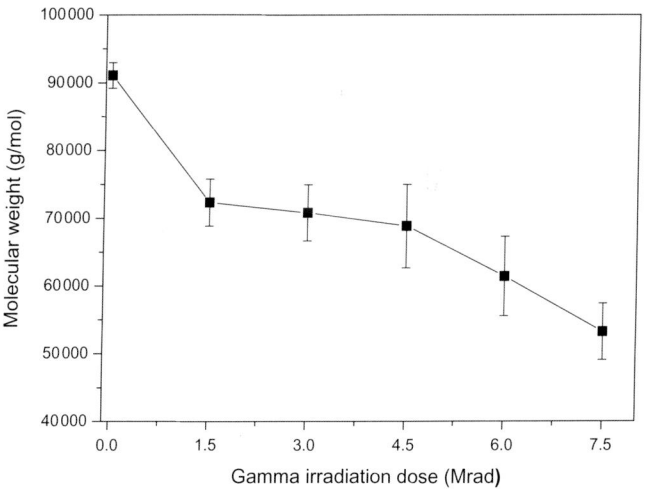

Fig. V.6 Changes of PLGA molecular weight according to gamma irradiation dosage.

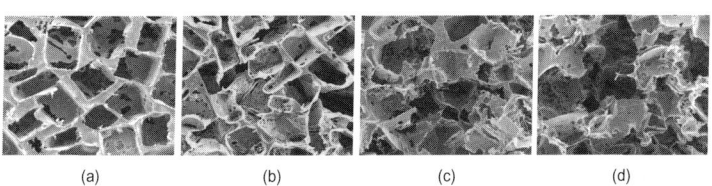

Fig. V.7 SEM microimage in cross-section of PLGA scaffold after gamma irradiation sterilisation. (a) 0 Mrad, (b) 1.5 Mrad, (c) 3 Mrad, and (d) 4.5 Mrad (magnification is 500).

V.4.3 Electron beam radiation

- Because of the issue of penetration distance, the use of electron beam radiation is much more limited. However, the availability of higher-energy/higher-power machines is lessening this limitation.
- Therefore, a unique application for this method is the in-line sterilisation of thin products immediately following primary packaging.

V.5 Sterility Test

- End medical products must pass sterility tests.
- The sterility assurance level (SAL) indicates the probability that a given medical product will remain nonsterile following exposure to a given sterilisation process.
- The generally accepted minimum SAL is 10^{-6}, i.e. one out of one million medical products will remain nonsterile after sterilisation.

Abbreviations

AMPEG	α-amino-ω-methoxy-PEG
BDNF	brain-derived neurotrophic factor
bFGF	basic fibroblast growth factor
BMP	bone morphogenic protein
CAD	computer-aided design
CATE	computer-aided tissue engineering
^{60}Co	cobalt-60
DI	deionised water
DMEM	Dulbecco's modified Eagle medium
DSC	differential scanning calorimetry
EC	ethylene chlorohydrin
ECM	extracellular matrix
EDC	1-ethyl-3-(3-dimethylaminopropyl)-carbodimide hydrochloride
EDTA	ethylene diamine tetraacetic acid
EOG	ethylene oxide gas
ePTFE	expanded polytetrafluoroethylene, Gore-Tex®
EthD-1	ethidium homodimer
FBS	fetal bovine serum
FDM	fused deposition modelling
FTIR	Fourier transform infrared spectroscopy
GA	glutaraldehyde
GPC	gel permeation chromatography

HBGF	heparin-binding growth factor
HEPES	N-2-hydroxyethylpiperazine-N-2-ethane sulfonic acid
HFP	1,1,1,3,3,3-hexafluoro-2-propanol
HGF	hepatocyte growth factor
HPLC	high performance liquid chromatography
IEP	isoelectric point
IGF	insulin-like growth factor
IleOEt	L-isoleucine ethyl ester
IPA	isopropyl alcohol
μ-CT	micro-computerised tomography
MTT	3-(4,5-dimethylthiazol-2-yl)-2,5-diphenyl tetrazolium bromide
MWCO	molecular weight cut-off
NGF	nerve growth factor
NHS	N-hydroxy succinimide
NMR	nuclear magnetic resonance
PBS	phosphate buffered saline
PBT	polybutylene terephthalate
PCL	poly(ε-caprolactone)
PDGF	platelet-derived growth factor
PDMS	polydimethylsiloxane
PDO	polydioxanone
PE	polyethylene
PED	precise extruding deposition
PEG	polyethylene glycol
PET	polyethylene terephthalate, Dacron®
PGA	polyglycolide
PHEMA	polyhydroxyethylmethacrylate
PLA	polylactide
PLCL	poly(L-lactide-*co*-ε-caprolactone)
PLGA	poly(lactide-*co*-glycolide)
PLLA	poly(L-lactide)
PMMA	polymethylmetacylate
PNIPAAm	poly(N-isopropylacrylamide)
PP	polypropylene
PPy	polypyrrole
PS	polystyrene
PU	polyurethane
PVA	polyvinylalcohol

PVDF	polyvinylidenefluoride
RP	rapid prototyping
RT-PCR	reverse transcription polymerase chain reaction
SAL	sterility assurance level
SEM	scanning electron microscopy
SIS	small intestine submucosa
SPCL	blend of starch with poly(ε-caprolactone)
TE	tissue engineering
TEA	triethylamine
TGA	thioglycolic acid
TGF-β	transforming growth factor-β
TGF-β_1	transforming growth factor-β_1
THF	tetrahydrofurane
TIPS	thermally induced phase separation
VEGF	vascular endothelial growth factor

Index